The Establishment of Science in America

The Establishment of Science in America

150 YEARS OF THE AMERICAN ASSOCIATION FOR THE ADVANCEMENT OF SCIENCE

SALLY GREGORY KOHLSTEDT

MICHAEL M. SOKAL

BRUCE V. LEWENSTEIN

Rutgers University Press

New Brunswick, New Jersey, and London

Library of Congress Cataloging-in-Publication Data

Kohlstedt, Sally Gregory, 1943–
    The establishment of science in America : 150 years of the
American Association for the Advancement of Science / Sally Gregory
Kohlstedt, Michael M. Sokal, Bruce V. Lewenstein.
        p. cm.
    Includes bibliographical references and index.
    ISBN 0-8135-2705-8 (alk. paper)
    1. American Association for the Advancement of Science—History.
I. Sokal, Michael M. (Michael Mark), 1945– . II. Lewenstein,
Bruce V. III. American Association for the Advancement of Science.
IV. Title.
Q11.K66 1999
506'.073—dc21   99-14096
                                                                                    CIP

British Cataloging-in-Publication data for this book is available from the British Library

Manufactured in the United States of America

CONTENTS

FOREWORD

STEPHEN JAY GOULD

Let me begin with an honest, if embarrassing, admission. I did volunteer to write this foreword, but I also felt internally compelled as a sometime historian of science, a man who loves to write (and fancies that he can), and as the president-elect of this august organization. But I had no intention of reading the book beyond a skim and a scan for some illustrative material. After all, as a literary genre, what could possibly rank below the commissioned institutional biography, that ultimate domain of hagiographical hyping for people and triumphalist trumpeting for organizations? These books are always ghostwritten for nonliterary bosses by authors who cash their paychecks and otherwise remain anonymous. Pride goeth before a fall. (The only worse genre, I suppose, is my current task of writing introductions to institutional biographies—for what could be less appealing than boilerplate upon hagiography?)

So I dunked my toes into this text, as planned, and then a funny thing happened. I read the whole book, every word of it, in one sitting, because the account was so informative, interesting, and honorably written. I really shouldn't have been surprised, and for two good reasons. First, the nonanonymous authors are all professional historians of science with excellent reputations for writing literate prose. No profession has worked harder to overcome the hagiographical treatment of science, once canonical in the history of our discipline (Dr. Arrowsmith marching on to victory and human salvation). Second, three authors treating three different and sequential periods implies three somewhat different concerns, approaches, and styles—thus adding the interest of diversity in the form of a mild Rashomon.

This book coheres in its variety by subtly commingling the two great themes that make history both instructive and interesting: the flow of *directional narrative* to tell a story, and the persistence of *general themes* to grant coherence—or the arrows and cycles, the unique configurations and the immanent generalities of our standard and paired metaphors about time. We feel the narrative pulse of a progressive history as the American Association for the Advancement of Science grows from a small club of geologists and naturalists in 1848, to the preeminent professional organization of all American science today (with membership well in excess of 100,000; a set of directorates with large staffs working in all major areas of interaction between science and society; a flagship publication of highest respect, quality, and circulation in the profession of science; and a new home in a stunning piece of modern architecture in downtown Washington, D.C.). But we also sense the perennial nature of issues that have troubled the organization from its inception, and that never got solved, but only recycled as the social context changes (how shall science be made more interesting and comprehensible to the public; how shall we balance the different claims and virtues of professional expertise in narrow specializations vs. welcoming breadth of interdisciplinary understanding?) Since the best scientists must rank as reasonably intelligent, even wise, people, I think we have to assume that an absence of solutions in 150 years of trying must imply a certain inherent intractability.

Three different writers; three different themes; three different times. Sally Gregory Kohlstedt treats the first fifty years, focusing on the nature of growth pangs, organizational identification, and the fascinating issue of establishing authority for a populist organization in a fairly élite profession within a democratic land. (A 1897 editorial in *Science* magazine actually suggested, more than metaphorically, that the more elite and self-perpetuating National Academy of Sciences should serve as an "upper house" of American science, while the more open and democratic AAAS should constitute the "lower house.") Michael M. Sokal treats the middle years of more entrenched stodginess, when the organization acted more as an umbrella for affiliated professional societies than as a platform for a coherent discipline, and when James McKeen Cattell played the increasingly less comfortable role (for the organization) of both owner and editor of *Science* magazine for more than fifty years. The baton of authorship then passes to Bruce V. Lewenstein, and to the very different set of growth pangs that led the AAAS from this rather uncontroversial lethargy of midlife into the maelstrom of an uneasy greater age in the post–World War II world of big science, military might, and political protest—not to mention all the unpredictable Armageddons of our forthcoming millennium.

Finally, through this narration, and through this cycling of recurrent and insoluble issues, we are also treated to the third, and last, essential element of all good stories: the *juicy details* of arresting incidents and interesting people (W.E.B. Du Bois as the first prominent black member; Thomas Edison and Alexander Graham Bell as early and failed owners of *Science*). In this final category of intriguing tidbits, consider the acrimony of a debate on cosmic rays at the 1932 annual meeting, after which Compton and Millikan, the two Nobel gladiators, refused to shake hands; or Warren Weaver's interesting idea for interdisciplinary outreach, when he persuaded the New York Philharmonic Orchestra to run speeches by scientists during intermissions of their radio broadcasts; or this remarkable tale, from the 1955 annual meeting in Atlanta, on the utter illogicality (not to mention the immorality) of racism: "Detlev Bronk, a recent president of both the AAAS and of Johns Hopkins University, stood in the rain in front of a downtown Atlanta hotel, furious because he could not get a taxi to take him to an AAAS session at the black Atlanta University. The white taxis could not take him to a black neighborhood and the black taxis could not pick up a white passenger."

Coherent and flowing narrative; anchoring and instructive generality; juicy and intriguing details. To these three components of good stories, we should add, in conclusion, a fourth reason that the history of the AAAS stands so far above the expectations of its genre; an organization consistently led by people of goodwill (whatever their inevitable and innumerable human failings), on behalf of a central institution in human life and history—the great adventure of science.

ACKNOWLEDGMENTS

This volume owes its immediate origins to a commission issued by Nan Broadbent, director of the AAAS News and Information Office, on the recommendation of the AAAS History Advisory Committee, appointed by the association's board as its 150th anniversary was approaching. Chaired by Keith R. Benson (University of Washington), the committee's other members include Bruce Hevly (University of Washington), Daniel J. Kevles (California Institute of Technology), Jane Maienschein (Arizona State University), Kenneth R. Manning (Massachusetts Institute of Technology), Gregg Mitman (University of Oklahoma), Naomi Oreskes (University of California, San Diego), Philip J. Pauly (Rutgers University), Nathan Reingold (Smithsonian Institution), and Jeffrey L. Sturchio (Merck & Company, Inc.). Nan Broadbent and the staff of her office (notably Dave Amber and Ellen Cooper) provided important administrative and logistical support, and many members of the committee carefully read and provided insightful reviews of successive chapter drafts. The committee also had drafts read by senior members of the current AAAS staff, including Richard Getzinger (director, International Programs), Shirley Malcom (director, Education and Human Resources Programs), Michael Spinella (director, Membership and Meetings), Michael Strauss (Meetings Program manager), and Albert Teich (director, Science and Policy Programs). The committee also arranged an especially thoughtful review of these chapters' penultimate drafts by Margaret W. Rossiter (Cornell University). The authors wish to thank these individuals for their support and, especially, for their comments and reviews, which have undoubtedly improved each chapter and the volume as a whole. The authors remain fully responsible for all content and appreciate the total autonomy under which they prepared their texts.

The authors—all historians of science—approached their assignments from perspectives deriving from long-term (in two cases, of almost thirty years) professional interests in the history of the American Association for the Advancement of Science. All three authors have thus benefited greatly from the continuous access to the Association's records and assistance with their use provided by long-term AAAS archivist Michele L. Aldrich (now associate editor of *Isis*, currently at Cornell University). Before Dr. Aldrich assumed control of these records, access to then-unarchived materials was provided by (among others) Catherine Borras (long-term executive office secretary), William Carey (executive officer, 1975–1987), Patricia Curlin (who served in several AAAS offices), Stacey Newton (archival intern), and Hans Nussbaum (AAAS business manager for more than thirty years). Since Dr. Aldrich's departure, oversight of the archives she organized devolved first to the AAAS Office of News and Information and then to the AAAS Directorate for Science and Policy Programs, whose directors and staff members have proved most helpful. In particular, as Amy Crumpton performed parallel research in support of the AAAS's exciting 150th anniversary exhibit (organized by Dr. Teich), she and the authors often worked synergistically and effectively supported the others' efforts.

As the preceding paragraph suggests, the authors examined the AAAS's internal records over a period of almost thirty years. Since the AAAS remained a working organization throughout these decades, the structure of its files and archives has continually evolved over time. The authors have made some effort to have their chapters' citations to this archival material reflect this evolution. They must admit, however, that their notes remain imperfect, and individuals seeking particular documents should be ready to explore the archives creatively.

As might be expected, as they began writing this book, each author took prime responsibility for the period of the AAAS's history with which he or she was most familiar. Sally Gregory Kohlstedt thus focused on the association's early years, Michael M. Sokal on its middle period, and Bruce Lewenstein on the more recent past. But as they prepared abstracts of their chapters and then began circulating chapter drafts, a real collaboration emerged from which all three authors have profited greatly. Although the authors prepared their chapters individually, each chapter embodies various insights provided by the prime authors of the other chapters—including the cross-referencing that integrates the whole—and each of the authors owes the other two a significant intellectual debt.

In addition to the appreciation already expressed, each author wishes to offer his or her particular acknowledgments, as follows:

Sally Gregory Kohlstedt's long-standing interest in the history of the AAAS received early support from Nathan Reingold and Marc Rothenberg, the successive editors of the Joseph Henry Papers (Smithsonian Institution), as well as the staff of the Smithsonian Institution Archives. Over several years her studies of the AAAS have benefited from the work of various graduate assistants at Syracuse University and the University of Minnesota—including David Sloane, Horace Taft, Mark Jorgensen, and Juan Ilerbaig—who have identified and collected relevant materials. In addition to those noted earlier, Michele Aldrich provided particularly helpful comments on an early draft of this volume's first chapter. Themes within this chapter reflect Dr. Kohlstedt's earlier work, including: *The Formation of the American Scientific Community: The American Association for the Advancement of Science, 1848–1960* (Urbana: University of Illinois Press, 1976); "*Science*: The Struggle for Survival, 1880–1894," *Science* 208 (4 July 1980): 33–42; and "Power through Persuasion: The AAAS and the Federal Government in the 19th Century," presented as part of a 1991 symposium on "The AAAS in Public Affairs, 1848–1975."

Michael M. Sokal completed much of his research on the history of the AAAS in the Manuscript Division of the Library of Congress, where successive librarians—most notably Mary M. Wolfskill, now head of the Library's Manuscript Reading Room—greatly facilitated access to the two hundred boxes of the papers of James McKeen Cattell (many crammed with AAAS-related material) and to dozens of other relevant manuscript collections. Richard P. Quitadamo (Worcester Polytechnic Institute) performed important statistical research, and (in addition to those noted earlier) Michele Aldrich and Toby Appel (Yale University) provided especially thoughtful comments on earlier drafts of this volume's second chapter. Sections of this chapter derive (at least initially) from several of Dr. Sokal's previous publications and presentations, including: "*Science* and James McKeen Cattell," *Science* 209 (4 July 1980): 43–52; "From the Archives: Cattell and World War II Censorship," *Science, Technology, & Human Values*, vol. 10, no. 2, (spring 1985), pp. 24–27; and "*Science* and Public Policy, 1900–1945," presented as part of the 1991 symposium on "The AAAS in Public Affairs, 1848–1975."

Bruce V. Lewenstein's studies of the history of the AAAS have been supported by many people associated with the AAAS and its activities during the past half century. Dennis Flanagan, John Pfeiffer, Gerard Piel, E. G. Sherburne, and Dael Wolfle have generously given him interviews and access to records under their control. He is especially grateful to Philip H. Abelson (editor of *Science*, 1962–1984) and to Dr. Wolfle for providing detailed comments on early drafts of this volume's third chapter, and to Dr. Wolfle for sharing many years ago a copy of his memoirs while they were still in draft form. John C. Burnham also allowed him to read in manuscript his book on popular science, *How Superstition Won and Science Lost: Popularizing Science and Health in the United States* (Rutgers University Press, 1987). Dr. Lewenstein would also like to thank archivist Tom Rosenbaum at the Rockefeller Foundation. Earlier versions of portions of this volume's third chapter have appeared in Dr. Lewenstein's earlier writings, including: his doctoral dissertation, "Public Understanding of Science in the United States, 1945–1965" (History and Sociology of Science, University of Pennsylvania, 1987); "Was There Really a Popular Science 'Boom'?" *Science, Technology & Human Values*, vol. 12, no. 2, (spring 1987), pp. 29–41, and "The AAAS and the Public Understanding of Science," presented as part of the 1991 symposium on "The AAAS in Public Affairs, 1848–1975."

These acknowledgments undoubtedly omit the names of many others whose assistance and insights have benefited the authors during the past two decades. To these individuals the authors offer their most sincere thanks.

Finally, though many late-twentieth-century scholars disparage acknowledgments of personal support by spouses—arguing that they've become all too common, or are gratuitous, or simply have gone out of fashion—the three authors disagree. Indeed, as they stormed at times to meet successive deadlines in the face of serious distractions, each found such support invaluable. Their sincere expressions of thanks to David L. Kohlstedt, to Charlene Key Sokal, and to Claudia Voss Lewenstein are thus anything but routine or gratuitous.

The Establishment of Science in America

Introduction

AAAS Narrative History

Keith Benson and Jane Maienschein

A century and a half ago, eighty-seven prominent American men of science—most of them geologists and naturalists—met in Philadelphia to form the American Association for the Advancement of Science. Today, 145,000 men and women—members from around the world—continue the work these men began, though the association's activities and the context in which they exist would scarcely be recognizable to those original members. These dedicated scholars pursued research in 1848 wherever they could by exploring the natural world. They wandered, wondered, observed, cataloged, collected, hypothesized, tested, wrote, talked, and taught. Why? Advancing science was not easy in a culture in which scientific pursuit was not obviously or universally valued and in which there was only a small scientific brotherhood. Nor was the United States particularly important in the grand scheme of world activities. But they did manage to create the important and far-reaching international scientific organization that the AAAS has become today.

The decision by a few scientists to "resolve themselves" into a new scientific organization on the American continent was unlikely to capture much world attention. With a world population of around one billion, the United States was a bit player in the world's community, with a population roughly that of France.

France in early 1848 was experiencing a political and intellectual revolution. Demonstrations led to the king's abdication, and almost immediately similar calls for political revolution echoed throughout Europe. The universities and intellectuals reveled in these opportunities to frame a new world order, now based on new ideas of modernism and enlightenment, leading to an advancing of knowledge and culture.

In Asia, Great Britain continued its expansion and exploration following the recent opium wars, while other European nations battled with local groups seeking to reassert their control of Africa. The daring exploration of Antarctica and the North Pole ultimately failed, but efforts continued to stretch the limits of human eagerness to examine their natural surroundings. Transport of convicts from Britain to Botany Bay (New South Wales) resumed after a temporary hiatus, moving people around the globe and raising new questions and new opportunities for exploration.

While American geology was beginning to command European attention, for the most part European intellectuals paid little attention to American scientists, and American intellectuals perhaps even less. Most Americans were preoccupied with issues of regional expansion and interregional strife, economic development, and, after 1848, gold. The deep and persistent regional rivalries that ultimately led to the Civil War helped to maintain the country as a loose federation of states more than as a nation with strong central identity.

Against this background, the group of scientists gathered in Philadelphia felt the need to assert a national identity as American scientists, and to establish a national community to discuss scientific affairs and promote "the advancement of [American] science" throughout the states. The new association provided continuity and a place for commitment to reason and empirical study of the natural world in an American setting. Scientists, along with all other Americans, were soon torn apart by bitter debates surrounding the meaning and acceptability of slavery and, ultimately, by the Civil War that resulted. Charles Darwin's theories of evolution by the mechanism of natural selection challenged religious thinkers and scientists, as both groups struggled to endorse the underlying naturalism and appeal to reason, while offering "American" interpretations of the mechanisms that Darwin offered.

By the late nineteenth century, the romanticism that had dominated European thought, and had found expression in America as well, had largely given way to pragmatic realism, with materialism and rationalism in tow. Scientists sat enthralled at national meetings of the AAAS, watching as physicist Albert A. Michelson and chemist Edward W. Morley demonstrated an experiment that raised profound questions about the assumption of the existence of a stationary ether. This, in turn, had profound implications for the way humans think about themselves and their place in the world. Clearly, this was one way that science had assumed a larger role in the United States, and the AAAS, now nearly a half century old, had the opportunity to play a central role. Nevertheless, the AAAS did not thrive at this time. The growing recognition of the importance of science in the United States led to the development of many additional academic and scientific societies, all attempting to serve roles similar to the one that the AAAS had begun to play. In addition, since the AAAS lacked a clear organizational structure, it was not in a good position to respond to other scientific societies.

By the turn of the century, professional scientists from many young disciplines, especially those related to the life sciences, were contributing to the early stages of the Progressive era, promoting regulation of food, drugs, and environmental conservation with their expertise. Certainly, the growing nation required more science and advanced the role of scientists. As James McKeen Cattell noted in 1925, "The advancement of science should be the chief concern of a nation that would conserve and increase the welfare of its people," a view that figured increasingly in the early twentieth century. Thus, the AAAS provided an organization and a community of scientists willing and able to step into the new roles and to engage in the use of science for public good.

World War I reinforced the new power of science and its technological companions, as new and impressive armaments and materials were developed for the military. This, as a number of writers at the time noted, was a golden age for science and engineering. Scientists could help to clean up the meatpacking industry, produce medical miracles, and provide automobile and even air travel possibilities for more and more Americans. As the 1920s roared with economic prosperity and its enthusiasm for modernism, scientists, doctors, and engineers gained prestige and support as the nation's new experts. In a world full of Elmer Gantrys, American novelist Sinclair Lewis chose a scientist—Martin Arrowsmith—as one of his few true heroes.

Much of the spread of science was the indirect result of the effects of the 1862 Morrill Land Grant program in bringing research universities to all the states of the Union, thus soon providing opportunities for young Americans to study at home instead of traveling to Europe for science. Additionally the immigration fueled technological growth beginning with the end of the century, and the political, economic, and religious struggles in Europe in the new century began to bring scientists to this country of expanding opportunity. These increasing roles for scientists, combined with the devastating effects of World War I in making it necessary for scientists to leave the old world, the worldwide economic depression, and Nazi policies of persecution, intensified the move by intellectuals, including scientists, to the United States.

Nevertheless, Americans had their struggles with science. Following World War I, there was a public reaction against all things European, especially German, and science sometimes became suspect. For example, while evolution theory was generally well received in the nineteenth century, the debate in the twentieth century took place in the context of the social concern, primarily religious, about evolution and atheism. The reaction to evolution and atheism was not necessarily tied to reaction against things European. The Scopes trial in the 1920s had become a challenge to the authority of science, while, at the same time, support continued to gather in recognition of advances in medicine, and pressure was exerted for establishing national research centers such as the National Institutes of Health.

Throughout this period, the AAAS had a voice in *Science* magazine, for though the association did not own the publication, its owner and publisher, James McKeen Cattell, had insured its financial success and the centrality of his own outspoken and self-confident voice by allying the magazine with this large scientific organization. This provided a forum for the "Declaration of Intellectual Freedom" in 1933, arguing for academic freedom and the end of suppression of Jewish scientists. Indeed, it was felt, that science could and should serve as a voice of reason.

World War II brought challenges to the sometimes questionable assumption that science remained pure and based on reason, and that science could serve the public good. On one hand, scientists helped in the war effort and the ultimate defeat of Adolf Hitler. But the means of the victory led the atom bomb physicist J. Robert Oppenheimer to declare that "scientists have known sin." Perhaps, critics began to suggest, the social contract between science and society needed to be renegotiated.

To that end, and in search of its own appropriate role in the changing world, the AAAS held a summit meeting at Columbia University's Arden House in 1951. Participants could reflect on President Truman's words at the AAAS annual meeting in 1948: "Now and in the years ahead, we need more than anything else the honest and uncompromising common-sense of science. Science means a method of thought. That method is characterized by open-mindedness, honesty, perseverance, and, above all, by an unflinching passion for knowledge and truth. When more of the peoples of the world have learned the ways of thought of the scientist, we shall have better reason to expect lasting peace and a fuller life for all." A few years later, Truman established the National Science Foundation. The United States soon moved into the space race and an era of renewed prosperity, with a central role for science and scientists and all their values—or so it seemed.

Senator Joseph McCarthy did not trust intellectuals, including scientists, and took every opportunity to challenge their credentials. Oppenheimer and other prominent scientists suffered considerably in this time of Red-baiting, but many others also felt the threats to academic freedom and to their integrity as members of society. The AAAS felt the need to respond, and the Arden House meeting brought a serious commitment to define the purposes and direction for what had become one of two dominant American scientific organizations. Alongside the National Academy of Sciences and its associated National Research Council, with their quite different mandates and memberships, the AAAS committed itself to advancing the enormous postwar growth in the U.S. research system, to serving on the front lines of the battle for freedom for scientists and their scientific ideas, and to engaging in the use of science for social good.

With the social unrest and challenges to authority of the 1960s and 1970s, the AAAS redefined itself again as the organization taking the lead in promoting science in the public interest. New education programs designed to improve the opportunities and access to science for all; activities addressing scientific policy and the role of the federal government in science; examination of issues of human freedom—both for scientists, and based on scientific study that showed fundamental equality of men and women and of all races; programs focusing on work with the media and on exploration of new media formats (such as the popular news magazine *Science 8X* or recent electronic publishing enterprises) intended to improve the understanding of science; efforts to improve the way that science plays out in law; exploration of the relationship of science and religion; documentation of changing funding patterns for science: all have been developed as core programs for the redefined and expanded AAAS.

A committed professional staff works in an environmentally progressive new building in Washington to carry out these programs, and to take the lead for scientists internationally as well as to respond to changing needs of the scientific community and the way it works with the larger society. The world's largest federation of scientific societies and individual members has become the leader in advancing science and its social roles, so that *Science* and the annual meeting attract lively interest. President Clinton and Supreme Court justice Stephen Breyer both addressed the 150th anniversary meeting in Philadelphia, for example, and *Science* remains one of the most cited science publications in the world. It would seem that the founders' original dedication in 1848 to the advancement of science has had rich results.

Recent years have brought what some might see as increasing public criticism and a dimming of science's luster. Yet this is, in fact, a developing appreciation of science as a human endeavor carried out as a process that is not inexorably progressive or infallible, but rather self-correcting and creative. The community sets standards and tests for science, while advancing opportunities for exploration and expansion. The AAAS has promoted improvement of scientific literacy for all, as a creative process based on research and active learning, conforming to a dynamic set of rules, and operating in a complex global and social context. The AAAS has not tried to hide the controversies and confusions by insisting on the purity and absoluteness of objectivity in science; rather, the organization has embraced the challenge to social commitment and has developed programs and hired leaders to implement change. Some have asked why the AAAS offers a program called Dialogue between Science and Religion, for example. What, these critics ask, does religion have to do with science? That is what the program seeks to explore, and by embracing the questions and addressing them rather than closing them out, the AAAS offers a powerful opportunity to develop synergies between the values of science and those of the larger society.

The 150 years of the AAAS reveal the exciting ways that science has advanced an understanding of the world while contributing to, and benefiting from, advancements in the diverse society from which science comes and that science enriches. This organization has always allowed science to remain central and important, while increasingly appreciating the social value and the importance of the public understanding of science.

This volume explores the emergence of the organization as we know it today, documenting the challenges and changes along the way. Three distinguished historians of American science have written fresh and exciting narratives to examine the three half centuries of the AAAS's 150-year history. Sally Gregory Kohlstedt has detailed how a small and ragtag group of naturalists and geologists worked together to develop a community of American science. Michael M. Sokal has written about the importance of the strong personality of James McKeen Cattell, who helped to develop a coherent administrative structure and messages for American science through his leadership in AAAS, especially as longtime editor of *Science*. Finally, Bruce V. Lewenstein has focused on the emergence of the AAAS as this country's leading institution for the public understanding of science in recent years. Together, these three essays provide a rich look at this important force in shaping American science and its place in the changing world.

CHAPTER 1

Creating a Forum for Science: AAAS in the Nineteenth Century

SALLY GREGORY KOHLSTEDT

On the morning of September 20, 1848, scientific leaders met in the library of the Academy of Natural Sciences in Philadelphia, excited by the prospect before them. They were about to create the first genuinely national scientific organization in the maturing United States and had every reason to believe that the moment held symbolic and practical significance as they voted to accept a carefully constructed constitution for the organization to be known as the American Association for the Advancement of Science. Interest was sufficiently high that the Academy could not hold all the prospective members planning to present and listen to papers. They reconvened in rooms at the nearby University of Pennsylvania for days filled with sessions on natural history and natural philosophy and, in the evenings, enjoyed hospitality proffered by local scientific and political leaders. News of the successful meeting circulated widely, establishing the AAAS as a national forum for scientific discussion.

An Institutional Foundation for American Scientists

The history of the American Association for the Advancement of Science from its founding in 1848 to the end of the century is, in large measure, an account of how creative and well-positioned scientists established scientific authority in the United States. They were self-conscious about the international stature of American science and eager to build institutions within which scientists could concentrate on research activities. A small cadre of educated men, concentrated primarily in eastern cities, built upon national and international precedents to create an organization that could coordinate and represent their ambitions for science in its economically expanding, avowedly egalitarian, and rapidly professionalizing social and political environment. Among the issues that they would confront were several that would recur throughout the association's ensuing 150 years: how to maintain an open yet focused forum for scientists to exchange information about their research; how to foster communication among disparate scientific areas of study, an outlook that might today be identified as interdisciplinarity; and how to embody the principles of democracy and popular access while advancing scientific knowledge that seemed increasingly esoteric to many people. As they shaped the AAAS to address these issues, the founders established an organization that provided a stable, public forum through which the scientific community could define, indeed continuously redefine, itself.

Their initiative was timely. The 1840s coincided with the maturation of natural history survey projects in several states. The return of the United States Exploring Expedition under Charles Wilkes brought subsequent discussion about the disposition of the wonderful natural history specimens collected by his and other expeditions. The Smithsonian Institution provided a visible new center for science in the nation's capital. Small colleges increased their curricular and informal activities in science. The public enthusiasm for science was evident in attendance at lecture halls, and sales of a wide array of periodicals, as well as of handbooks and textbooks designed for the teaching of particular sciences. Not surprisingly, then, the founders drew on broad public sponsorship as they launched their new scientific flagship organization. Its auspicious beginnings could not preclude some rough seas during the Civil War, and its hegemony would be challenged both by more prestigious institutions like the National Academy of Sciences and, later, by the specialized societies representing chemistry, physics, and an array of biological interests. Its steady course through this turbulence, however, maintained the fundamental aspirations of the leading scientists, who took their turns as officers and thus created the most visible center for North American science in the nineteenth century.[1] Inevitably, the AAAS became a forum not only for presentation of new scientific knowledge but also for debate about how science should establish and maintain itself in American life.

It was, in many ways, a century of science, in Europe and throughout its imperial networks. Scientists engaged in institution building, in clarifying methodology, in constituting global networks, in developing comprehensive taxonomies, and in increasing instrumentation deemed essential for scientific research. During the first half of the century, both natural and physical sciences were pursued by individuals whose most important contacts were through local societies or private correspondence, although the popular taste for science had been growing rapidly through popular periodicals and public lectures. For the last half of the century the AAAS thus became a central, indeed singular "living" reflection of shifts in scientific self-definition, and its institutional configuration and publications complemented private correspondence, public discussions, and widely circulated essays on the nature of science and its practice.

Among the founding leadership, no goal was more important than overcoming geographic distances separating scientists and creating a coherent voice and cohesive center for scientists. The fundamental, indeed essential, goal was to make the AAAS a viable and highly visible forum for science. Its annual meetings were intended to develop direct professional connections among the scientists while simultaneously presenting a public display of scientific achievement, and the subsequent *Proceedings* would constitute its permanent record of activity and outreach. Meetings from Montreal to Charleston and from Philadelphia to Cleveland in the decade before the Civil War made clear the national aspirations of the AAAS. A second aspiration, often introduced by negative references to "pseudo-science" in the presidential lectures of the 1850s, was to establish and disseminate "standards"—a term that might refer to the quality of research, uniformity in method and measurement, or ways of creating peer rather than public evaluation of scientific work.[2] A third purpose, particularly for the Washington-based AAAS leaders, was public advocacy in support of science and particular scientific projects. The mechanisms of annual meetings, publications, and committees for activism became much more elaborate over time. In the twentieth century, projects relating to dissemination, education, and diversity came to play a much larger role. The three formative goals of definition, coordination, and advocacy, refined and redefined, persisted over the course of the century. By the end of the century, however, centrifugal forces of disciplinary definition and specialization challenged the original vision of a cohesive, singular scientific organization and required a rethinking of its functions and establishment of fresh strategies to retain a genuinely national and scientifically comprehensive organization.

Coordinating Scientists

Although an emphasis on individualism permeated antebellum America during and after the era of President Andrew Jackson, the growing cadre of scientists understood that much of their work required cooperation and well-orchestrated collaboration. Natural history and natural philosophy had often been somewhat solitary pursuits, enhanced for most by private correspondence and occasional visits, aside from the few fortunate to live in seaboard cities. This individual activity starting in the colonial period had perhaps provided unusual opportunities for women like Jane Colden of New York, who had corresponded with and contributed American specimens to Carl Linnaeus, and for others not part of the social mainstream, like Benjamin Banneker, whose African American and Native American family heritage did not prevent him from becoming a respected instrument maker and almanac writer in the late eighteenth century.[3] Reacting to the theme of isolation, however, various efforts, some quite successful, had been made during the early years of the republic to form societies in the natural sciences.[4]

Perhaps no group was more keenly conscious of the need for teamwork on research and for coordination among parallel investigations than the geologists. Seeking to map their topography and identify natural resources, a number of state legislatures established geological and natural history surveys. Since geological formations have scant relationship to political boundaries, cooperation was essential as theoretical ideas about mountain building and formulations of stratigraphy were debated among geologists and in final reports presented to sponsoring state agencies. One obvious mechanism was an organization of geologists that would establish a focused scientific agenda, recruit beyond the local constituencies represented in other natural history and philosophical societies of the time, and address a truly national set of problems. It was soon evident that some geologists had even greater aspirations for the organization, echoing sentiments expressed by other naturalists and physical scientists throughout the 1830s. Within a decade, there was a genuinely broad-based and comprehensive national scientific association.[5]

The Initiative of Geologists and Naturalists

Working geologists and naturalists who had found full-time and long-term employment in science were keen to enhance government patronage and general public awareness of their results.[6] Their material and scientific circumstances positioned such natural scientists well to undertake a national scientific organization. Beginning with the North Carolina survey in 1821, eastern seaboard states initiated surveys, initially quite modest but growing in some instances to employ several geologists, zoologists, and botanists, to evaluate their natural resources and topography. In 1836, the New York and Pennsylvania legislatures authorized significant projects, and they attracted some of the leading scientists in the young nation. A few, like Henry Darwin Rogers, had attended meetings of the very successful Geological Society of London, and others had presented papers before the British Association for the Advancement of Science. In creating the American Society of Geologists in 1840 (the term "Naturalists" was added in 1842 to broaden the membership base), prominent and well-connected geologists like Rogers, Benjamin Silliman, Sr., who edited the nation's only comprehensive *American Journal of Sciences*, and Amherst professor Edward Hitchcock intended to sponsor an annual meeting for scientific discussion and a publication to reflect the quality and results of geological and natural history research conducted on the surveys and under private auspices. Deeply conscious of their tenuous intellectual status as former colonials, and frustrated that current philanthropic and government practices did not provide much sustained support for science, the members of the AAGN focused on making their new organization creditable on the international as well as national scene.[7]

Not wanting to compete with other natural history societies, the geologists emphasized the importance of having a place where survey scientists could learn from each other's research on contiguous terrain and parallel theoretical questions. New York geologists initiated the first meeting in 1840 but held it at the well-positioned Franklin Institute in Philadelphia, involving rustics like Douglass Houghton from as far west as Michigan. The AAGN's attendance and membership expanded quickly; at the 1842 meeting in Boston, there were forty members, and a crowd of five hundred people attended an evening address by chemist and mineralogist Benjamin Silliman, Sr., of Yale. The eminent geologist Charles Lyell, who had been touring North American geological sites, attended the sessions as well as formal dinners and pronounced the gathering a success. Equally important, two leading geologists, Henry Darwin Rogers of the New Jersey and Pennsylvania surveys, and his older brother William Barton Rogers of the Virginia survey, unveiled their path-breaking, if controversial, theory "On the Physical Structure of the Appalachian Chain, as Exemplified by the Laws Which Have Regulated the Elevation of Great Mountains Generally," subsequently published in the subsidized first volume of *Reports* by the Association.[8]

Some measure of the effect of the young organization was suggested by William Barton Rogers, who enthusiastically wrote to a colleague, "For us such reunions of the scientific brethren as our Association of Geologists are of precious value and form the best compensation we can enjoy for the prolonged restraints of our vocation. What new impulses to exertion, what encouragement and guidance do they not give? and then in our hours of lonely meditation how many cheering and delightful *social* recollections. . . . "[9] Even when the group dynamic included intense debates, such lively exchange underscored the importance and the identification of issues that might only be resolved through additional research. Solidly established and evidently serving constituent members well, the AAGN became an obvious base on which to build a more comprehensive national scientific society not only "of" a group of scientists but avowedly "for" the advancement of science more generally. For those geologists and naturalists, "advancement" meant acquiring better intellectual and material resources for their work.

A Broad Base for the Scientific Community

Key scientific leaders facilitated the transformation of the AAGN into the American Association for the Advancement of Science in 1848. In fact, the fundamental decisions had already been made at the 1847 meeting of the AAGN, led by its own set of leading geologists and a number of prominent visitors representing other scientific fields. The presence of Joseph Henry, the Princeton physicist who the previous year had been named secretary of the new Smithsonian Institution, was an important signal of the impending decision. The 1847 meeting also hosted Swiss naturalist Louis Agassiz, then on a highly popular lecture tour in the United States and soon to be named professor at Harvard University. Several Harvard faculty members, including mathematician Benjamin Peirce, also showed considerable interest in helping shape a national scientific organization that would incorporate some characteristics of the well-attended and publicly acclaimed British Association for the Advancement of Science. Agassiz and Peirce joined Henry Darwin Rogers in writing the "Rules and Objectives of the Association." The Association's intentions were clearly stated: "By periodical and migratory meetings, to promote intercourse between those who are cultivating science in different parts of the United States, to give a stronger and more general impulse, and a more systematic direction to scientific research in our country; and to procure for the labours of scientific men, increased facilities and a wider usefulness."[10]

Founders wanted the unprecedented, truly national society to be organized in a way that would support existing local societies engaged in natural history or in broadly defined arts and sciences by its occasional presence, strengthening rather than competing with them. At a celebratory meeting that opened on the morning of September 20, 1848, in the library of the Academy of Natural Sciences of Philadelphia, the AAAS was launched; and by the end of the meeting 461 members were on the rolls. The ensuing patterns of participation and cosponsorship suggest that goals of inclusivity across geographic boundaries and scientific fields were, in fact, being met.

Somewhat more complicated was the issue of individual participation. While the BAAS had, by 1848, established several distinct categories of membership, and sharply limited voting privileges to a general committee composed of officers, publishing members, officers of other scientific societies, specially designated delegates from such societies, and recommended visiting foreigners, the AAAS created no firm categorical distinctions.[11] Henry Darwin Rogers explained in a letter to his brother William that he sought to make the AAAS constitution "Democratic, federal, flexible and expansive, progressive, with all the true conservatism these features imply."[12] William was in agreement that an open membership policy would, in fact, counterbalance any arbitrary leadership, maintaining steady and thus "conservative" course for the new organization. While his brother was in Britain, William presented the document to the 1848 meeting, where it was approved as the AAGN metamorphosed into the AAAS.

Somewhat ironically, these prominent geologists found themselves gradually pushed outside the emerging leadership circle. By the mid-1850s, William Barton Rogers had become a counterweight, and he spearheaded an "opposition" when the new leadership seemed to exercise its authority arbitrarily and to form a largely self-perpetuating and thus—to him—"anti-democratic" Standing Committee. While he fended off the most apparent and arbitrary abuses of power by a group of scientists who privately called themselves the Lazzaroni (the name selected for an inner circle who believed the name fit their posture as scientific "beggars"), the elder Rogers brother largely turned his attention elsewhere and during the next decade was a major architect in creating what became the Massachusetts Institute of Technology.[13]

Assuming central authority, both in visible ways and behind the scenes, was a group of scientists whose connections to Alexander Dallas Bache, "chief" of the Lazzaroni, made them sympathetic to his aggressive agenda of promoting a new scientific leadership and positioning it to gain more financial support for science. As head of the United States Coast Survey, Bache was in a powerful position to advance the careers of younger scientists, and, as a great-grandson of Benjamin Franklin, well connected to the Philadelphia elite, he had strong alliances with political leaders in Congress.[14] In his outgoing presidential address to the AAAS in 1851, he observed that, "where science is without organization, it is without power."[15] For him, then, the intellectual agenda envisioned by the AAGN founders was important perhaps primarily as a means for the organization to gain authority and for that authority to be based on the scientists' status as experts. Associations, Bache argued, should foster the best possible scientific presentations in the hope they might stimulate "second-rank" scientists, but they should also provide a mechanism for weeding out incompetents, charlatans, quacks, and pseudo-scientists. A decided advocate for the objectives that would come to charac-terize science after the Civil War, Bache stressed the need for greater support for scientists, higher standards for research, and more attention to "abstract research," and even hinted at the possibility of creating a more highly selec-tive scientific organization. His message resonated well among good friends like Joseph Henry and Benjamin Peirce but also attracted an ambitious set of younger scientists like astronomer Benjamin A. Gould, who also worked for the Coast Survey. It is important to note that, in fact, the Rogers brothers and Bache agreed on many aspects of scientific organization. Committed to emerging ideals of professionalism, all three actively promoted government sponsorship of science, better collegiate and even postgraduate education in science, a code of scientific ethics, and some distinction between amateurs and professional scientists. Personality and philosophy, however, distanced them considerably. William Barton Rogers, the optimistic son of an immi-grant Irish physician, was committed to a largely open, democratically oper-ating organization, while the more calculating Bache was clearly embedded in the upper-class tradition of Philadelphia in his emphasis on leadership elites and private management of organizations. Other emerging scientific leaders, like the young Spencer F. Baird, who was in charge of the zoological collec-tions of the Smithsonian Institution and served briefly as secretary of the AAAS, preferred not to become involved in these internal debates.

Membership and Leadership in the 1850s

Guided by Bache and his colleagues, the AAAS became a highly vis-ible, coherent, and cohesive organization, one that could effectively "aid in bringing together and combining the labours of individuals who are widely scattered, into an institution that will represent the whole." The leaders also intended to use the AAAS, through its annual meetings and publications, as a forum for offering advice on scientific questions and to argue for support of scientific projects. During this first decade, a core group of about twelve men from the north and mid-Atlantic seaboard states, who held various of-fices and presented papers at the annual meetings, contributed to and largely directed AAAS affairs. These included Agassiz, Bache, Henry, Peirce, the Rogers brothers, as well as Benjamin A. Gould, geologist James Dwight Dana, chemist Oliver Wolcott Gibbs, meteorologist William Redfield, and chemist and geologist Benjamin Silliman Jr. During the period from 1848 to 1860, a much larger group of over three hundred people presented at least one paper or held at least one office; and the published lists of members establish the fact that over two thousand people joined (many of them local citizens who only came for one year but by their attendance indicated the attraction of the public scientific meetings.[16] Available membership records make clear the correlation between the location of the annual meeting and the recruiting of new members into the AAAS, suggesting that this young organization conveyed a certain status and opportunity even for those local supporters whose interests were less participatory than observational. Although less active, these local members helped AAAS leaders define their scientific community.

Like its two-year-old predecessor, the Smithsonian Institution, the AAAS was committed to the increase and diffusion of knowledge, providing the latter primarily through gala annual meetings. Following the precedent of the British Association, the AAAS members found themselves feted in most of the host cities, provided reduced fares on the railroad lines when attending meetings, and generally lionized by the newspapers and periodicals that reported on their scientific talks and social events. It was usual for local scientific societies to invite the AAAS to their libraries, museums, botanic gardens, and other scientific facilities, and typically a significant number of their members joined the national association during the occasion. Wealthy or prominent local citizens and sometimes a city council member would provide homes and funds to host one or more evening receptions; civic-minded duty led some to sponsor science as well as the arts. When the association met in Newport, for example, they were invited to Julia Ward Howe's nearby estate. Some Southerners like Joseph LeConte declined the invitation because Howe was so outspokenly against slavery and in favor of women's suffrage.[17] Sometimes, too, prominent citizens like ex-president Millard Fillmore of Buffalo would host the truly distinguished guests during the week of meetings.[18] At the conclusion of each conference, an elaborate vote of thanks, recorded in the *Proceedings*, reviewed details of local hospitality proffered the visiting savants. In some cases, the local sponsors also raised money to help publish the papers from the meeting, apparently wanting to have their city featured prominently on the title page of the *Proceedings*.

The AAAS leaders typically enjoyed the annual festivities and sought to accommodate those newspapers willing to devote long columns relating to the scientific sessions in their daily publications. The quietly composed Quaker astronomer Maria Mitchell observed wryly, "For a few days Science reigns supreme—we are feted and complimented to the top of our bent, and although complimenters and complimented must feel that it is only a sort of theatrical performance for a few days and over, one does enjoy acting the part of greatness for a while."[19] Another member quipped to a *New York Times* newspaper reporter in 1858, "Think of it! Men who have been sleeping on rocks, eating out cold victuals off of logs, digging in the earth, and breaking stone nine months of the year—bringing us to soirees as these and making us (unintentionally of course) laughing stocks."[20]

The issue of newspaper reporting was always complicated. The AAAS was committed to a public display of scientists and their work, and dissemination through the press brought the scientific results before a public much larger than those in attendance. But all too often the reports by nonscientists were "imperfect" or "erroneous" and concentrated on a few "eccentric individuals" who furnished a "fruitful theme for the reporters' wit."[21] On another level, too, the press understood the value of sensational stories and particularly liked to report on intense debate or on accounts of highly controversial topics. Thus the press covered a somewhat ad hoc discussion on the unity versus plurality of "the races" during the Charleston meeting in 1850. Samuel George Morton had argued, based on his study of skulls, that there were innate differences in human types, and his former student Josiah C. Nott of Mobile, Alabama, presented supporting arguments on the purity of the "Jewish race" to the Charleston members. Louis Agassiz weighed in, as well, on the side of plurality of races and was quoted at length in local newspapers and soon thereafter published his views in the *Christian Examiner* (July 1850). The southern audience was receptive, given the effort of many to argue that African slaves were a distinct race, but Agassiz by his open arguments positioned himself in opposition to some of his Cambridge colleagues, and his ideas would also put him in opposition to Darwin's evolutionary ideas in *Origin of Species.*[22] None of this controversy, including the extended comments by Louis Agassiz reported in local newspapers, was published in the official AAAS *Proceedings.*

At some meetings a hired stenographer took notes with the goal of providing more accurate reports, but this comprehensive and often tedious reporting had its own downside. The pioneering science reporter John Swinton of the *New York Times* commented, for example, that "there is a real hungering and thirsting after science in the popular mind; but the savans [sic] are silent or write only for their brethren."[23] The complaint about inaccessibility of scientific information would even become stronger in later decades. The association could effectively enhance the visibility and reputation of scientists through media coverage, but the written reports did not always meet heightened public expectations about better understanding of scientific activity and results.

The peripatetic AAAS held meetings in places calculated to reinforce such well-established scientific organizations as the Boston Society of Natural History and the Franklin Institute of Philadelphia, and even to reinvigorate or create new institutions. One success story was the reopening of the natural history museum in Charleston, with the subsequent formation of the Elliott Society of Natural History there in 1853.[24] Often scientists who did not make an annual trek to the AAAS meetings would attend those held in their region, strengthening their ties not only to prominent national scientists but also to the regional scientific elite with whom they were more likely to have continuing contact. This coordinating function seems particularly true in the antebellum South, where, for example, plantation owner and leading American mycologist William Henry Ravenel made important connections during the 1850 meeting in Charleston.[25] In Albany in 1856, the AAAS meeting orchestrated by state geologist James Hall involved the dedication of two major scientific facilities in New York's capital, his geological museum and the Dudley Observatory.[26] In a subsequent turn of events, a controversy over appointments to the Observatory made Albany civic leaders less enthusiastic about outside experts and the AAAS leaders somewhat more cautious about intruding in local enterprises. The downside of an AAAS visit could also be to demonstrate the weakness and perhaps contribute to the demise of a local society, like the Western Society of Cincinnati, whose members voted not to host any local events collectively and to attend the AAAS meeting as individuals in 1853.[27]

For enthusiasts of the AAAS, a recurring theme was that while formal meetings were important, "in our own land, as well as in Germany, England and France . . . by far the greatest good is done outside the walls of the section-rooms." In these informal meetings, as Benjamin A. Gould continued in his 1868 address, "ideas exchanged, suggestions offered, asperities softened, co-operations established . . . the legitimate and constant fruits of these meetings."[28] Adding to the social and informal quality of the meetings were the excursions to local museums, colleges, and significant natural sights like Niagara Falls. Some were systematic field trips intended to demonstrate a significant scientific site to colleagues, while other jaunts allowed members and their companions to explore new places and experience similar institutions in unfamiliar places. In later years, annual meetings were enhanced with displays of books and instruments, particularly after an impressive demonstration of historic and current models of microscopes at the Salem meeting in 1869.

The process of shaping the AAAS's institutional infrastructure and negotiating the complicated questions around the standards for membership and publication inevitably led to controversy and even disillusionment and withdrawal by a few during the first decade. Many of these controversies were shaped by competing ideas about the meaning of democracy, an idea that always held more meaning than that of simple political participation. Tension that would never be fully resolved arose around complicated issues of membership credentials and scientific methods, as its recurrence after the Civil War demonstrated. Debates recorded primarily in private letters and public newspapers reveal the ambivalence of antebellum Americans toward issues of elitism versus democracy and strong central authority versus the needs of subspecialties, echoing in some ways the issues of federalism and states' rights that plunged the nation into the Civil War. By providing the single best forum for discussion of scientific issues, the AAAS leaders were occasionally embarrassed; but, overall, the organization demonstrated its usefulness and resiliency. Although the number of members dropped in the immediate post–Civil War years, it climbed again until the late 1880s and early 1890s, when membership reached a plateau and participation in the annual meetings again slipped into decline. (See Appendix C.)

Defining Scientists in Principle and in Practice

The self-conscious new leaders of the AAAS were intent on making it an organization through which individual credentials, research results, and scientific standards (including methods and presentations) could be established. The midcentury scientific initiative coincided with the professionalizing efforts of engineers, physicians, and lawyers. While the scientists did not seek the formal licensing being established by the more narrowly defined occupational groups, they evidently hoped to provide mechanisms by which they and the public could evaluate individuals and their work. The very term "scientist" had only recently been introduced by the English natural philosopher William Whewell, and important indicators of scientific status revolved around membership in peer organizations and publications. Within the association itself, working definitions of AAAS membership would be revised several times over the first three decades.

Addresses of outgoing presidents, given before a large public and often printed, became both a means of presenting new scientific work or reviewing recent accomplishments in the field and also, in many cases, an opportunity to talk about the state of science and the importance of scientific standards, including an increasing emphasis on method. The National Academy of Sciences, established in 1863 with just fifty of arguably the leading scientists in the country, seemed to signal a quite different style of scientific culture and one more similar to that in European nations with similar organizations. Amid criticism of the self-selecting and tightly controlled membership, tension continued to surround the process of professionalization inside the AAAS.[29] The question of standards and procedures for membership became particularly intense (and much more public) within the AAAS because the organization had included people who were themselves positioned at different points at the boundaries among popular enthusiasts, practicing scientists, exceptional researchers, and scientific administrators. Boundary disputes occurred at the interstices between scientific enthusiasts and practicing scientists over who might present and publish papers, for example, and negotiations persisted over questions about the managers of particular scientific sections, who might be more or less flexible about presentations and discussion. The territory claimed by emerging professionals, however, was rarely yielded again.

Bache and His Agenda for Science

Close colleagues Joseph Henry and Alexander Dallas Bache self-consciously positioned themselves to speak for science, and the AAAS provided one important public podium. Bache presented, in 1851, the first major public outline of professional aspirations, goals that resonated particularly well with the younger generation of physical and natural scientists who would become leaders in American science during the last half of the century. The previous year, in an address that was not published at the time for reasons that Nathan Reingold speculates may have been personal but equally likely were strategic, Joseph Henry as outgoing president had already suggested the role of such lectures:[30]

They should present to the public, from time to time, the character, the doings, the past history, and the present condition of the Association. They should vindicate the claims of science to public respect and encouragement and set forth the nature and dignity of the pursuit; [they should] give instruction from past experience in the method of making new acquisitions from nature and point out new objects and new paths of research; they should expose the wiles of the pretender and suggest the means of diminishing the impediments to more rapid scientific progress . . . they should call attention to the relation of science to the moral part of our nature and . . . point out all things which may have a bearing through science on the material and spiritual improvement of man.[31]

While Henry's own address did not accomplish all these tasks, he made clear the problems that good science might face and argued for a workable code of ethics. Although asked to devise such ethical guidelines, Henry must have found the assignment difficult, and he never presented a report to his AAAS colleagues.

Bache's presidential address the next year was carefully timed. The AAAS was firmly established, had a growing membership, and had gained public attention at home and abroad. This was the moment to forge consensus around two matters that concerned him, namely the relationship between amateur and professional scientists as well as the relationship between what would later be designated basic and applied research. Seeking to establish a higher reputation for American science abroad, Bache suggested that "charlatanism" was a significant threat, ranging from outright quackery to the ungrounded opinions of those who had expertise in one area presuming to hold an opinion in another area of science. Whiggish by upbringing, Bache supported a relatively open membership but wanted decision-making to be managed through appointed committees and the standing committee (forerunner of today's board of directors), on which he served for the remainder of the decade. He thus supported critical review of all papers to be published in the annual volume of *Proceedings*, as well as a role for the organizing committee to determine which papers might be presented as a means of insuring the quality and nature of the science on display.

Engaged in the empirical project of gathering extensive data on shorelines and tides, Bache was convinced that the work was based on and would contribute to fundamental scientific questions. Like others educated at West Point, he believed that basic science was antecedent to and companion of applied science and insisted that the AAAS emphasize the importance of systematic support for abstract research as well as empirical studies. Moreover, in his 1851 AAAS address he suggested that at some future point the government should provide an organization similar to the French Académie des Sciences, which provided financial support to outstanding researchers and had sufficient resources to publish scientific monographs. This address was often referenced by peers, marking it as a milestone in the formulation of guidelines and definitions by which the nascent scientific community might operate.

Members and Standards

During its founding decade, the incipient forms of review by peers were made part of program planning and conference publications in the AAAS. While the Royal Society of London and other well-established scientific societies also consulted with peers before accepting a publication, the review procedures were now (intermittently) applied to a much more democratically conceived and open operating association. The result could be highly controversial, as in the case of the *Proceedings* from the annual meeting in Cleveland in 1853, when the local committee published all papers submitted, only to find its version suppressed and withdrawn from distribution, to be replaced by a version that had passed the scrutiny of the standing committee.[32] The tensions ran so high over this seemingly arbitrary decision that some of the geologists seriously considered withdrawing from the AAAS and forming another organization more in keeping with the collegial relations in the much smaller AAGN.

Not long after this controversy, the powerful standing committee sought to create more stringent membership criteria, and their initiative brought a strong reaction among some members of the association. From the outset, the AAAS welcomed essentially everyone who expressed an interest in science, requiring no scientific credentials, and thus included doctors and lawyers, teachers and clergymen, and an array of other amateurs and cultivators of science.[33] To many AAAS members this very diversity of membership defined the scientific community. The rosters from the 1850s included a range of individuals from transcendentalist writer Henry David Thoreau, to former president Millard Fillmore, and to Maria Mitchell, famous for her discovery of a comet and for receiving a medal from the king of Denmark.[34] Her selection to membership in 1850 made moot the occasional question raised about women members, and by 1860 there were others, including Eunice Foote and the well-known scientific textbook author Almira Hart Lincoln Phelps. Many more women would join in the decades following the Civil War.[35] Attitudes about race appear to have been more intractable, despite a few exceptional early inventor/inventors like Benjamin Banneker; and there is no evidence of active African-American scientists in the AAAS during the nineteenth century.

During the nineteenth century, attitudes toward disability were changing, however, and the deaf were encouraged to study science; when Joseph Henry spoke at what is today Gallaudet College in 1876, he argued in terms reflecting contemporary opinion that deaf people "are specially well adapted to various [unspecified] scientific investigations." As a result of access to education, a number of deaf scientists were active members of the AAAS, including paleobotanist Leo Lesquereux and Gerald McCarthy, state botanist and entomologist of North Carolina.[36]

Although the membership lists are not necessarily indicators of actual attendance, the fact that so many diverse individuals accepted or sought membership, even temporarily, indicates the importance accorded science by midcentury. For the most part, there were few complications, as the active scientists were keen to make their ideas known, and most amateurs were content to listen quietly. This was not invariably true, however, and the exceptions demonstrate the more limited sense of community that some of the elite leaders held. While the *Proceedings* almost never recounted problems, newspapers occasionally revealed complaints registered by men whose papers had not been accepted or even by those whose presentations had been interrupted by the chair of the session, as at the Cleveland meeting.

Not long after the controversy over the 1853 Cleveland meeting, the powerful standing committee sought to limit full membership by imposing more stringent criteria and creating a category of "associate member." Their action brought a strong reaction from some members of the association, especially those not consistently active in scientific pursuits. Although the new category was formally adopted in a constitution of 1857, the hotly contested nomenclature that was intended to distinguish among members was, in fact, not implemented, perhaps because even the leaders were aware of the complexity of making fine distinctions in a period when an outstanding physicist like Joseph Henry lacked a college degree. Another outcome, however, was that the negative feelings generated by the recommendation of the standing committee led William Barton Rogers and a few others largely to withdraw from AAAS affairs and to turn their attention elsewhere. At the same time, however, a significant number of practicing engineers joined, and their papers on applied topics were readily published. Some, like Joseph G. Totten, a chief engineer in the U.S. Army who was associated with various military and public works projects along the Atlantic coast, had both technical and scientific interests; in Totten's case, he used his assignments along the shoreline to became an expert on shells.

By far the most serious disruption for scientific activity generally and, more specifically for the AAAS, was the Civil War. In 1860, well aware of sectional tensions and hoping to demonstrate that science was not subject to regional and national barriers, AAAS leaders deliberately selected Nashville, Tennessee, for the 1861 meeting. After war broke out and quickly involved even the border states, however, the meeting had to be postponed indefinitely.

Private correspondence from the war years reveals some continuing reflection about the future of the association. Some of its members were the scientific sponsors of the new, more highly selective National Academy of Sciences, established in 1863. While a few scientists may have viewed the NAS as a replacement, a much larger group wanted to revive the AAAS after the Civil War and, in some sense, to acknowledge the complementary functions of the two groups. The NAS, after all, had only a limited membership and hoped to focus on joint research done to assist in government activities; the AAAS had already established its capacity to draw public attention and to promote scientific activity. Moreover, the NAS would be dominated by physical scientists; the AAAS continued to represent all the emerging disciplines well. Correspondence reveals that many older AAAS members missed the camaraderie of the annual meetings and agreed with William Barton Rogers's observation: "Whatever may have been its faults, its shortcomings, it was of great service, and its free & unexclusive spirit ought I think to be perpetuated."[37]

Frederick A. P. Barnard, the last president elected before the Civil War and newly selected president of Columbia University, initiated the revival. The 1866 meeting was convened in Buffalo in the week following the NAS annual meeting in Northampton, Massachusetts, in order to allow members of both organizations to attend. The AAAS meeting in upstate New York was a quiet, collaborative affair of a group sobered and changed by the war and by other circumstances. Joseph Lovering, the Harvard-based permanent secretary, soon found ways to increase revenue and stabilize membership, instituting a one-time initiation fee and tightening the nomination process to require signatures of current members; and he also sought to revitalize committee activities. Bache died shortly after the war ended, and the older Joseph Henry, demoralized by a major fire in the Smithsonian, was less active than before. Southern science did not recover quickly, and individual scientists did not have the means or the incentives to rejoin their Northern colleagues in significant numbers.[38] In subsequent years, the association continued to meet in smaller cities, consolidating its strength among the growing number of scientists in state agencies, private industries, and colleges, as well as those in well-established federal agencies and in the Eastern seaboard cities.

Hierarchies and Distinctions

No demographic study has been done of post–Civil War AAAS membership, but such work would likely reveal tension existing between the persisting leadership of the largely male, Protestant, Anglo-Saxon officers and the pluralism of the larger society (and even the membership itself)— contributing in subtle ways to the persisting issues of status and participation. In the early 1870s, the standing committee reopened the possibility of formal distinctions among members and moved to create a higher, more prestigious category of "Fellows" for members "professionally engaged in science" or those who "by their labors aided in advancing science."[39] Although Robert E. Rogers, William Barton Rogers's brother, protested this elitism, the change to the constitution seems to have been passed with little real challenge. In the same year as the new constitution went into effect, 1874, the AAAS also became officially incorporated under a Massachusetts charter. The new category did help attract back some old members, including Southerner Joseph LeConte (who had, in fact, moved to a faculty position at the University of California at Berkeley after the Civil War), who was made a fellow and resumed paying dues regularly.[40] This olive branch to Southerners, however, may have helped exclude African Americans as members, since there is no evidence of such members until W.E.B. Du Bois joined in 1900.

As Margaret Rossiter has shown, this designation of "fellow" created a two-tiered system that frequently became a signifier of the differential status of men and women in science in a number of scientific organizations in the 1870s and 1880s. While the number of women members in the AAAS continued to increase during this same period, only a small number (and percentage) were ever named fellows. Rossiter observes that in the late nineteenth century, the very term "professional" was in some contexts "a synonym for an all-masculine and so high-status organization."[41] The AAAS never closed its doors to diversity and amateurs, as did other organizations, but it found ways to declare itself democratic while establishing an elite, thus negotiating the equity and access some of the earliest leaders had envisioned. At the same time, it is important to note that there were always some male leaders who actually encouraged women's membership, including Permanent Secretary Frederic W. Putnam. In fact, it was his colleague Alice C. Fletcher, a long-standing member who had earlier served on a committee regarding archaeological remains on public lands, who as the head of the anthropology section was the first woman to serve as head of an AAAS scientific section in 1896.[42]

The fellows list expanded rapidly in the 1880s. Scientific sections had by then been established, six in 1870 and a revised set of nine in 1881, to accommodate growing and changing special interests of members; and each section was allowed three nominations each year. Later changes in the constitution also required that elected officers be fellows, thus creating a permanent hierarchy, supplemented by the even more distinguished category of "honorary fellow." By the end of the century, the number of fellows was only slightly less than the number of members (in 1900, 1,052 members and 865 fellows). The movement toward establishing distinctions had been met by practices that continuously enlarged opportunities, so that the AAAS never lost its image as an open organization, albeit one that acknowledged individual achievements.

Method and Substance in Science

The most prominent scientists willingly served as presidents and delivered addresses intended to inspire and motivate their audiences. Some, like the often cited address of Alexis Caswell in 1859, documented the current state of a particular field—in his case, astronomy—underscoring the excellent instrumentation available and the need to provide better institutional and staff infrastructure to make the field internationally recognized.[43] In the post–Civil War years, American scientists often decried the absence of what they perceived as more powerful and vital organizations abroad, but all of them recognized that the annual AAAS meetings remained the single most effective vehicle for discussing the nature of science and for delineating its rapidly shifting boundaries. John Strong Newberry's address in 1867 set the tone for the post–Civil War period, a time when contention made people uneasy and scientists hoped to refocus the AAAS on scientific issues rather than structural problems. While the AAAS forum operated at several levels, the special addresses by the president, and in later years by the vice presidents of the scientific sections, were occasions for identifying and addressing scientific or public concerns. Not all were published, and not all seem, in retrospect, particularly timely or timeless. But some of them were among the more important public statements of the time on science, and they helped contemporaries focus on important issues. They remain useful for those of us seeking to understand science more than a century later.

Asa Gray, American's leading botanist and an early supporter of Charles Darwin's evolutionary theory, used his presidency as an opportunity to revisit major research themes. The 1872 meeting was initially planned for San Francisco but was changed to Dubuque, Iowa, when the AAAS could not negotiate free or discounted train rates to California for East Coast members. Gray nonetheless went to California, where he visited Yosemite Valley in the company of colleagues from the new University of California at Berkeley, together with naturalist writer John Muir. Gray's extended trip allowed him to explore firsthand the species that, originally studied by him in the herbarium, suggested close relationships among flora of southern and eastern North America and of eastern Asia that were distinct from those of western North America. What he saw reinforced his important arguments about the migration and isolation of species. On his return trip across North America, he addressed the AAAS members in Iowa on "The Sequoia and Its History" in a presidential talk that marked a climax to his long career and his staunch support of Darwin and evolutionary theory. He retired that same year from the Harvard faculty.[44]

Gray's discussion of evolution at the 1872 meeting was more indirect than direct. It fit a pattern suggested later in a retrospective presidential address by Edward S. Morse, in 1887. Morse, who had reviewed publications in zoology relating to evolution since Darwin's *Origin*, commented, "The papers were first tinged with the new doctrine, then saturated, and now, without reference to the theory, Derivation is taken for granted."[45] He also credited Darwin for having created widespread interest in zoology, and he outlined the important work done on habits, mimicry, the balance of conditions between organisms, and the influence of environment. By the early 1870s, aspects of evolutionary research and theorizing had found their way into the AAAS in much the way described by Morse.

Some former Harvard students of Louis Agassiz, including Alpheus S. Packard, Frederic W. Putnam, and Morse himself, presented papers to the AAAS on classification, morphology, and embryology, as did Edward D. Cope, Burton G. Wilder, and others. The limited explicit discussion of evolution in this early period came from a few scientists like geologist George C. Swallow and botanist Thomas Meehan. After years as a geologist in Missouri, Kansas, and Montana, Swallow had recently been appointed professor of natural sciences at the University of Missouri. In 1873, he presented a paper arguing that there were two significant barriers to evolutionary processes: that between plants and animals, and that between lower animals and humans.[46] Meehan, trained as a horticulturist at the Royal Botanical Gardens at Kew, in London, ran a nursery in Germantown, Pennsylvania, edited *Gardener's Monthly*, and was botanist for the Pennsylvania State Board of Agriculture. He read widely, published extensively, and in 1874 presented a paper to the AAAS arguing that his botanical work supported evolution, but with significant markers rather than as a gradual process.[47] Their papers were published with apparently little comment.

In 1875, when entomologist John Lawrence LeConte was president and Canadian geologist John W. Dawson was vice president of the section on natu-

ral history and biology, members of the AAAS heard presentations from leading scientists directly on the issue of evolution. LeConte, the leading expert
on coleoptera in the country, offered evidence from his research on the distribution of that order against evolution, while Dawson drew on paleontology to reject theories of evolution based on the continual existence of the
same forms through time. The following year, Edward Morse followed
Dawson in the vice presidency of section B (natural history and biology) and
offered a more positive review of the literature by Americans on evolution.
Perhaps his role as director of the Peabody Museum in Salem made him particularly aware of public understanding, and his address was intended to make
more transparent the technical work of colleagues so that their contributions
could be better understood by those outside the field. For Morse and many
American contemporaries, evolution had a neo-Lamarckian orientation, and
he was very much influenced by the director of the Boston Museum of Natural
History, Alpheus Hyatt. As Morse anticipated in his 1887 address, discussions of evolution remained more or less in evidence through the remainder
of the century, especially at the Boston meeting in 1898 as part of the "half
century" review of science in the AAAS.[48]

Some presidents took up larger public issues, as when Simon Newcomb's address on "The Course of Nature" tackled the complex issue of the
relationship between natural science and Christian religion. Attendance at the
St. Louis meeting in 1878 had been limited by a yellow fever scare and
the uncomfortable August heat, but, nonetheless, the presidential address ultimately reached "a large audience and induced a spirited reaction."[49] The
AAAS issued Newcomb's address as a separate pamphlet, and other periodicals like *Popular Science Monthly* also published a complete text. This speech
took the ground held by several leading American scientists in the 1870s,
arguing that science and religion were distinct, albeit complementary, enterprises. In so doing, the outspoken mathematical astronomer also separated
himself from both the more strident stands of lecturers like agnostic Thomas
Henry Huxley and the claims of natural theology that had been incorporated
in the AAAS address of his presidential predecessor Asa Gray, among others. Newcomb's central proposition was that "science concerns itself only with
phenomena and the relations which connect them, and would not take account of any questions which do not in some way admit of being brought to
the test of observation." Challenged by men on both sides of the issue,
Newcomb was embroiled in a debate taken up particularly vigorously by
prominent colleagues like Gray and James McCosh, the president of Princeton
and philosophical leader of Scottish realism in the United States, a position

that resisted any simple equation of science with atheism. While the AAAS provided a base where terms of the debate could be formulated, it did not become the ongoing forum for a discussion that moved largely into the periodical literature and chapels on college campuses. Science and religion would continue to negotiate their respective places in the hearts and minds of individual scientists as well as in the larger society.

Fundamental in a quite different way was the question of scientific practice, and the AAAS meeting provided a platform for updating the discussion of scientific participation and the status of science. At the 1883 meeting in Minneapolis, Henry Rowland's "Plea for Pure Science"—presented as an address before the physics section, like Bache's address more than thirty years earlier—struck a resonant note with at least some in his audience of aspiring physical scientists. Ironically, this National Academy of Sciences member found that he needed to use the AAAS public forum in order to make his point to both fellow research scientists and a broader public. Educated at Rensselaer Polytechnic Institute, the young physicist had been offered a position at the new Johns Hopkins University and had spent some time in Europe for additional study. As vice president for the physics section of the AAAS, he delivered an address that was at once a powerful indictment of the situation for scientists in North America and a call to action; it has also been interpreted by some historians as evidence of American indifference to basic science.[50] Still fresh from his experiences abroad, Rowland made it clear that a concentration of resources in European academies had pulled physics there to a high level of achievement.

As Daniel Kevles has suggested, the talk was really a plea for "best science elitism" that concentrated resources into the laboratories of the scientists most acknowledged by peers as outstanding.[51] Disparaging the over three hundred American institutions of higher learning—Rowland called them a "cloud of mosquitoes, instead of eagles"—he wanted to highlight the potential of new research-oriented universities like his own Johns Hopkins. Not only were American institutions inadequate, he claimed; the presumptions about what constituted science had become blurred into such vocations as teaching and commerce. More free to state his case baldly than Bache had been in the 1850s, Rowland observed, "It is a fact in nature, which no democracy can change, that men are *not* equal—that some have brains and some hands."[52] The former, those who required time and resources for the "true pursuit" of their science, were to be identified and supported. His address not only made evident the increasingly insistent distinction between professional and amateur, but also elaborated a scientific calibration between the methods of the physicists in their laboratories and those of the naturalists who sustained local societies. For Rowland, the salient reference point was Western Europe, and the major research agenda established there was to be pursued in the United States as well.

Advocating Science through the AAAS

The ideal of having the AAAS as an advocate for science was hardly original; the founding leaders anticipated that role from the outset. They quickly positioned their organization to be effective in the public arena, most particularly in matters of public policy. Whether recounting the role of scientific patrons in antiquity and the seventeenth century or simply pointing to the effective lobbying of the British association, American scientists argued that the AAGN and later the AAAS should use its collective expertise to advance science by gaining more sponsorship for it. From the 1850s onward, the AAAS played a significant role as arbiter but most often as advocate, frequently but not exclusively, on the federal level.

Early Patterns of Influence

Leaders of the young association were not hesitant to take action. A. Hunter Dupree's study of science in the federal government notes that the association began its advocacy of science within its first year, and that half of the committees undertaking investigations in the first five years directed their resolutions at the federal government; he also points out that scientists employed by the federal government often served on such committees.[53] Joseph Henry's private comment perhaps reflected the thinking of his activist peers: "If the scientific men of the country will only be properly united they can do much for the advance of their pursuits through assistance from Congress. . . . Politicians as a class are timid except when they are sure of an object which they know is worthy and in advocating of which they are sure of being sustained by authority."[54] Henry and other North Americans who had traveled abroad often spoke and wrote enviously of the success of their counterpart scientific organizations in England and Germany in gaining government patronage.

The early AAAS membership responded to a significant range of initiatives, some of which came from the standing committee and others that arose from concerns by rank-and-file members. One of the first resolutions was a request to the State Department to intervene on behalf of a foreign professor of astronomy whose work was disrupted by political problems in Schleswig-Holstein. The geologists continued their pattern of requesting endorsement for new and continuing projects, and, indeed, within its first five years, the AAAS assembly passed fourteen resolutions memorializing state legislatures on the advisability of particular projects, most typically geological surveys. The committees and their resolutions were circumspect and self-conscious in their advocacy, often emphasizing that the goal was the disinterested one to "give direction" and a "stronger impulse" to scientific inquiry. A memorial to the New York legislature in 1852, for example, reflected this sensibility: "The American Association for the Advancement of Science has hitherto had the good fortune to receive from the public bodies to which it has addressed itself, the most favorable regard. It is perfectly aware of the grave responsibility imposed upon it by these marks of respect and consideration, and is therefore careful not to utter sentiments and opinions, whether speaking through its committees or otherwise, which cannot bear the most rigid scrutiny. Under this sense of duty and responsibility the present committee is now acting."[55]

Perhaps it was due to the sense that too much lobbying would undermine credibility, the real shift of science to the federal level, or even a changing political climate in the years immediately preceding the Civil War, but the number of committees and resolutions intended to lobby state legislatures declined in the latter half of the 1850s and never again reached the high point of the early 1850s. Federal lobbying, moreover, never became as extensive or influential as that of the British association during the nineteenth century.

The most consistently mentioned federal agency was the United States Coast Survey. An "evaluation" committee was established in 1849 and was activated whenever Bache believed that there would be some opposition to his annual appropriation in Congress. The observant Maria Mitchell noted, "The leaders make it pay pretty well. My friend, Professor Bache, makes the annual meetings the opportunity for working sundry little wheels, pulleys and levers; the result of all which is that he gets his enormous appropriations of $400,000 out of the Congress every winter. . . ."[56] There is little doubt of the close relationship between these committees and the Survey itself. When a concerted attack on the Coast Survey was launched in the press and in Congress in 1858, Bache's side was ready. They wrote an eighty-eight-page report produced under Frederick A. P. Barnard, who became influential in the revitalization of the AAAS after the Civil War. Published separately in 1858 and in the *Proceedings* in 1860, the report bears the unmistakable imprint of Bache's organization and praises the Survey's fiscal economy, its utilitarian benefits to commerce, and its international scientific standing. The appropriation passed readily.

Other federal agencies also benefited from high visibility at the annual meetings and occasional endorsement for their activities, including the Nautical Almanac, the Naval Observatory, and the Corps of Topographical Engineers. The AAAS also regularly petitioned Congress to insure that scientists would be included on all new exploring and surveying expeditions given federal sponsorship. These AAAS committees, appointed by the president or the standing committee, were subject to the same oversight as were publications and presentations; a few had their recommendations published separately in pamphlet form.

Individual members could also bring scientifically related issues before the AAAS membership, to provide publicity for certain subjects as well as to gain the imprimatur of the AAAS. Thus, the venerable naturalist Chester Dewey in 1856 spoke of the imminent extinction of giant fir trees in California before a AAAS general assembly and proposed a resolution to "save the residue of these giant inhabitants of our primitive forests from destruction." In this case Joseph Henry was asked to correspond with appropriate authorities in California and Washington, D.C., and he made a report on forests and their economic uses one of the early projects of his administration in the Smithsonian Institution. Conservation issues generated some local controversy and later debate in Congress but the broad-based argument about attending to natural resources made it a common good for most of the scientists and public. Thus, the AAAS would weigh in on the side of conservation throughout its history, without much discussion about the growing distinction between those who supported wilderness preservation and those concerned about wise and fair use of natural resources. Evidence in the AAAS *Proceedings* indicates that vigilance needed to be perpetual, and the AAAS after the Civil War passed resolutions relating to conservation topics, from use of Western ranges to preservation of the giant sequoia, throughout the remainder of the century. Woven into the advocacy were other AAAS concerns about coordination and the definition of community, leading to resolutions that involved the easy exchange of books and instruments across national boundaries without the imposition of tariff and taxes.[57]

Post–Civil War Policy Initiatives
Among the most active early advocates of forest preservation was Franklin R. Hough, whose concerns were aroused by data gathered during his term as superintendent of the 1870 United States census. The results led him to present a paper on "The Duty of Governments in the Preservation of Forests" at the 1873 annual meeting. Also prominent was member William

Henry Brewer, whose pioneering lectures on tour and at Yale's Sheffield Scientific School made him an evident leader for an AAAS Committee on the Preservation of the Forests, established in 1874. Although a bill to set up a commissioner of forestry in the Department of the Interior failed, a rider to another bill for the Department of Agriculture managed an appropriation of $2,000 to "prosecute investigations and inquiries" into forest problems. The purpose was to investigate scientifically the rate of forest consumption as well as to establish methods for renewal. Conservation historians argue that the AAAS support, when buttressed by popular writings promoting the same themes, was a critical component in Congressional action.[58] The committee persisted with a series of statutes and resolutions until 1880. In retrospective addresses later in the century, Bernhard E. Fernow, chief of the Forest Division, credited the AAAS with founding his division and noted its influence as well on the state forest policy initiatives in New York, California, and Georgia. Perhaps as a result, Fernow's own persistent reports in the early 1890s urged a more independent forest administration, a result accomplished under the more managerial leadership of Gifford Pinchot near the end of the century.[59]

Concerns about standardization and uniformity within the sciences in terms of nomenclature, measuring devices, instrumentation, and operating systems resulted in international discussion at various venues, including the occasional world fair and special congress in the nineteenth century. AAAS members followed the discussion of the metric system, in particular, beginning in 1850 with Henry Taylor's letter regarding a universal system of scales and standards for scientific measurement. A committee appointed to consider the problem submitted no report, but one member produced a pamphlet that addressed only the issue of international coinage. In 1855, Bache, because his responsibility for weights and measures under the U.S. Coast Survey (a responsibility it retained to a large degree until 1901) sensitized him to the issue, made a preliminary report; in 1860, he presented a progress report on the matter.[60] While no specific action was taken within the AAAS or nationally, the ongoing discussion positioned members for the more systematic, if unsuccessful, movement to implement the metric system after the Civil War. Then a bill was brought before Congress, and President Barnard proved a strong advocate during discussion at the AAAS meeting in 1866, but no formal resolutions were passed. In 1869, some AAAS members, in a similar spirit, endorsed the idea of a future meeting of the International Statistical Congress in the United States. The issue of standards and uniform nomenclature persisted, and in 1887, the AAAS supported a resolution in favor of a separate Bureau of Standards that could move the issue forward more effectively.[61]

The motivated committees of the 1850s were less in evidence, however. By the late 1880s, *Science* editorials reflected what seemed to be a lackluster performance by commenting, for example, that the reports were "in general as unsatisfactory this year as heretofore," and noting that some committees that had made no report were continued in the hopes that they might do so in the next year.[62] Those that did report, often verbally to the council, might mention conversations with members of Congress, reports of scientific observations, requests to add the name of the association to an invitation to hold an international congress in the United States, and other pro forma actions. Among the most disappointing initiatives were those intended to investigate and make recommendations on science education. Although a number of committees were appointed, few of them issued any substantial reports—and this stood in sharp contrast to the efforts and effects of the British association during this same period, spurred by Thomas Henry Huxley, in advocating more science in primary and secondary education.[63] Without direction from any AAAS administrator, the committees were only as strong as their most ardent members and sponsors; few, it seemed, had the requisite commitment or perhaps confidence that a report would have a larger impact.

In the absence of an organization for international science, AAAS members intermittently sought to build relationships with other national associations, particularly those in the British empire and later Commonwealth. Indeed, after the British association met in Canada in 1884, about three hundred members joined the AAAS meeting in Philadelphia, marking this the only meeting in the nineteenth century to register more than a thousand attendees. Under reciprocal agreements, an individual member of the AAAS might be made, for example, a temporary member of the Australasian Association for the Advancement of Science when it met in Christchurch, New Zealand, or of the BAAS when it met in Bristol. As transportation and communication continued to improve, there were also significant efforts to establish truly international standards not only of weights and measures but also of names and principles of organization for data, from astronomy to zoology.

Throughout the century, and particularly after the publication, in 1859, of Charles Darwin's *Origin of Species*, new principles of definition and arrangement of species created strong debates among zoologists about nomenclature, synonymy, and their relationship. The subgroup of entomologists was particularly persistent in its internal contentions and used the AAAS to try to resolve the complicated issue in a collective way.[64] Its members created an Entomology Club, affiliated with the AAAS, and persuaded the broad membership of the association to establish the Committee on Nomenclature in 1870. The debate had a temporary resolution when the Entomology Club voted during the AAAS meeting in Buffalo in 1877 to adopt an eleven-point set of rules that was accepted in slightly modified form by the entire AAAS assembly the following year, although conflict about the issue persisted until there was an international agreement on nomenclature. The association, with its comprehensive membership and expressed commitment to national scientific interests, was positioned to comment, and it did provide a forum for national specialists, who were then sometimes asked to consult on international methods and standards. Thus, for example, the AAAS created a Committee on Colors and Standard Nomenclature on Colors (1895) and a Committee to represent the Association as Members of the Advisory Board on International Codes of Zoological Nomenclature (1896).

The postwar years brought, in fact, many fundamental changes in America that had a profound impact on science. The federal government became increasingly large and powerful; colleges and universities positioned science and technology more centrally within the curriculum; concentrations of capital provided opportunities for scientific investment; and the new immigrant population of North America was being concentrated in cities, even as some immigrants spread westward. The AAAS reflected some of these changes as it concentrated its lobbying efforts more exclusively on the national government, worked to draw membership from land grant institutions across the country, and expanded its perimeter for annual meetings by visiting three cities west of the Mississippi River by the turn of the century.

Diffusion of Science

Popular interest in science, at least among those reading general circulation magazines, continued and expanded. The *North American Review* (1821), *Harper's Monthly Magazine* (1850), and Frank Leslie's *Popular Monthly* (1876) all regularly featured articles about science, especially after midcentury. Specialized magazines, too, like *Scientific American* (1845), helped create a market joined most prominently by Edward Livingston Youman's *Popular Science Monthly* (1872) and the new *Science* magazine (1880), underwritten by Thomas Edison and later by Alexander Graham Bell and his father-in-law, Gardiner Greene Hubbard.[65] An array of publications kept the public aware of prominent scientists and their activities, although the publications struggled for survival, and some failed. Such regularly published magazines upstaged the AAAS as a means of presenting highly current scientific news, but they also underscored the importance of the association's annual meeting by covering it thoroughly.

The stalwart William Barton Rogers, the founding president of MIT in the 1860s, welcomed the AAAS members to Boston and Cambridge in 1880, and attendance at the meeting soared as they returned to a major city. Rogers remained true to his early, open conceptualization of the association's purpose and accessibility: "But while we are an association for the advancement of science, we are in the very nature of the case instrumental in its diffusion, as is well shown by the comparatively popular character of the meetings of the American Association."[66] The following year, Rogers was named first honorary Fellow, perhaps an acknowledgment of his persistent vision and commitment to both advancing and diffusing science.[67]

Certainly the annual meetings continued to garner local press coverage as well as attention from newspapers with high and broad distribution like the *New York Times*. In an era before scientific news services and press secretaries, there was little effective translation of the scientific sessions into public idiom, however. Even reporters were frustrated by the changing complexity of science, leading a *New York Times* correspondent to report in 1875 that the younger scientists' presentations seemed more technical and less intelligible than those of their older peers. Commenting on the scientific presentations in Detroit, he dryly observed that members were "pouring out acre after acre of learned papers, whose weird interest and thrilling nature are finely illustrated by the fact that the morning papers which have printed them nearly in full, have lost a large portion of their already small circulation this week."[68] By the 1880s, the newspaper coverage had begun to fall off, perhaps, as *Science* put it, "in despair as to what to select from the superabundance of material."[69]

Certainly none of the scientists wanted to "dilute" the scientific sessions, but most supported the public evening sessions, and some participated in other popular venues as well. Spencer F. Baird, assistant and later secretary of the Smithsonian Institution, for example, both contributed such articles and served as a clearinghouse for colleagues who submitted news for the "Editor's Scientific Record" in *Harper's Monthly*. The AAAS meetings often figured in such summaries and editorial discussions about scientific activities and results. Moreover, the AAAS presidential addresses had with some regularity been published in the older *American Journal of Science* and in the much newer periodical published in Salem, *American Naturalist*. It was the advent of *Science* magazine in 1880 that provided a direct mechanism to coordinate coverage of science in ways that paralleled the AAAS. Thomas A. Edison, who was partial underwriter of the new enterprise, attended AAAS meetings to make connections with scientists. When in 1883 entomologist Samuel S. Scudder became editor of *Science*, which was by then supported by Alexander Graham Bell, he informed his colleagues, "The aim of the journal will be to increase the knowledge of our people, to show our transatlantic friends our real activity, to gain among intelligent people a knowledge of the true aims and purposes of science, and to elevate the standard of science among scientific men themselves."[70] Given the similarity of agendas between the AAAS and *Science*, it is not surprising that the magazine typically produced lists of presented papers, covered particular sessions, and published the presidential and other prominent addresses in the issues following each annual meeting, as well as issuing announcements about future meetings. The editorial board meetings for *Science* were often held in conjunction with the AAAS meetings, where the editor could also solicit news and articles. This ad hoc connection, encouraged more by some editors than by others, was formalized in 1895, when *Science* accepted a subsidy from the AAAS and created explicit mutual obligations codified by contract and providing for nominal AAAS oversight.

Research Funds within the AAAS

Historically, science has been primarily the prerogative of those wealthy enough to have the leisure to pursue it or those exceptional enough to find patrons. As this patronage moved into such institutional bases as academies, surveys, societies, observatories, and museums, the patterns of sponsorship became more varied as well. By mid-nineteenth century, the American government had developed an investment in science through the Corps of Engineers, the Coast Survey, and other agencies, while private sponsorship built instruments like telescopes, and supported local institutions. The "bounty for research" particularly benefited astronomy and natural history during the nineteenth century, and AAAS leaders were quite aware that the new association might potentially provide more than only moral support for projects.[71] At the time, however, the membership dues barely covered the cost of publishing the annual *Proceedings* and costs incidental to the annual meeting.

Models were available. The Rumford Fund of the American Academy of Arts and Sciences stood as an example of how private funds could provide incentives for original research, and the British Association for the Advancement of Science had acquired permanent research funds through donations, a fact to which Bache alluded in his 1851 address; by the mid-1880s the BAAS had nearly $8,000 available for research grants. Serious about the proposition but with an alternative beneficiary that more closely matched his vision of an elite scientific organization, at his death in 1867 Bache left over $50,000 for scientific research to the National Academy of Sciences.

It was not until a private but scientifically interested woman, Elizabeth Thompson, donated a gift of $1,000, in 1873, that the AAAS was in a position to advance scientific research with direct subsidies. One immediate result was the official incorporation of the AAAS (1874) in Massachusetts in order to "hold and convey real and personal property."[72] A childless wealthy widow from Boston, Thompson had donated money to a number of local charities and also claimed to follow popular reports of "every new discovery in science, every invention in mechanics." Her contribution was to be used to offset the "financial difficulties which beset those noble men of science who labor more for the truth than for profit's sake."[73] Almost immediately the funding was used to underwrite the expensive publication of Samuel H. Scudder's *Fossil Butterflies* (1875), with its wonderfully illustrated plates, produced as volume one of an AAAS *Memoir* series.

Once incorporated, the AAAS leaders could institute new mechanisms to create a permanent research fund. They established life memberships, hoping that an influx of fifty-dollar investments would help accumulate endowment money. No records of the AAAS survive that indicate clearly how funds were solicited, but the annual report on the treasury suggests nothing near the thousands of pounds that had by that time been accumulated by the BAAS in its research fund.

Then, a decade after her first benefaction, Thompson offered another contribution to be used to promote an International Scientific Congress under joint directorship of the AAAS and BAAS, hoping to encourage more transatlantic conversations. Interest in a formal connection had undoubtedly been stimulated by the decision of the BAAS to hold a meeting in Montreal in 1884, after which about three hundred British scientists went on to attend the Philadelphia meeting of the AAAS that had been postponed until September in order to make the dual attendance possible—by all accounts making both meetings unusually successful.[74] The proposed International Congress was eagerly discussed but never materialized, and the planning committee disbanded in 1887. Its prospective donor instead put $25,000 into an independent Elizabeth Thompson Science Fund administered by a private board of trustees.

Gradually, however, a research fund was established in the AAAS, and small grants in the range of $50 to $250 were awarded, based on income from investments. Applications and nominations for funds were handled by the sci-

ence sections as a form of peer review. One of the first went to Albert
A. Michelson and Edward W. Morley for their work on the velocity of light;
Michelson would later be the first American to win a Nobel Prize in science.
Another grant went to W. O. Atwater for his study of "heat combustion" in
vegetables. By 1897, there was about $6,000 in the research fund, and about
$600 was spent on research each year.[75]

The Allison Commission

The politics of late-nineteenth-century federally sponsored science were
dramatically encapsulated in the Allison Commission.[76] In 1884, Congress
appointed a commission headed by Senator William B. Allison of Iowa to
investigate agencies that were, or had been, among the most powerful scientific agencies in the federal government. Their significant government patronage (about $3 million in annual appropriations) came under scrutiny, even
attack, during a time when there were significant discussions of reorganization and various charges of ineptitude and corruption. Each of the named
agencies had its supporters and its enemies. Members of the National Academy of Sciences used the hearings as an opportunity to express their views
about scientific methods, sponsorship, and authority—but their ranks did not
hold steady despite their concentration in Washington, with independently
wealthy Alexander Agassiz the most notable defector.

The AAAS was not able to play the same kind of continuing role as
the Academy, which had close ties through the exclusive Cosmos Club in
Washington, D.C. By its 1885 meeting, however, AAAS members, reacting
to attacks in the press, decided to renew their long-standing support by voting "approval of the extent and high character of the work performed by the
U.S. Coast Survey." Without making explicit reference to Charles Sanders
Peirce, whose work on pendulum experiments had been challenged as an
inappropriate use of government funds, the AAAS resolutions asserted the
principle that "the value of the scientific work performed in the various departments can best be judged by scientific men." Echoed in this claim was
the theme of the former chief of the Coast Survey, Alexander Dallas Bache.
The AAAS was only one of a chorus of institutional voices, but its collective effort may well have been a factor in the resistance to the draconian cuts
and changes proposed.[77]

The Allison Commission, after nearly two years of heated debate in private and in the public press, took no action on its own. Eventually all the
corruption charges were dismissed. Both those who had hoped to check the
autonomy of the agencies in question and those who had planned to estab-

lish a strong central authority were also disappointed. The reform advocates who wanted to create clearer lines of jurisdictional authority were pleased with one outcome—Congress mandated that each bureau's charges for printing be listed separately and passed through the Appropriations Committee. This decision was symbolic and significant, given the fact that the federal government had been the largest publisher of scientific monographs in the preceding forty years. There is little evidence that printing requests ran into subsequent problems, and, by the 1880s, scientists were placing more emphasis on periodical publications and finding new venues for publications. The fact that there were such minimal changes suggests that legislators, heavily lobbied by members of various national scientific groups like the AAAS, basically reaffirmed the scope and purposes of government-sponsored science.

Holding the Scientific Center

While the history of the AAAS in the last third of the nineteenth century includes fewer dramatic moments and strong personalities than in its first decade, the role of the association as a stabilizing force should not be understated. Anthropologist Frederic Ward Putnam was instrumental in providing a kind of consistency and integrity in the midst of the turbulence in the larger political and scientific communities. He proved unable to present any new vision for the association during his long term in office as permanent secretary; he was fond of pointing out, instead, the association was, perhaps reluctantly, the "great mother organization of American associations of learning."

Putnam and the Establishment in Salem

Frederic Ward Putnam, with a New England pedigree that stood for tradition, was in fact crucial to the persistence and stability—and perhaps also to a particular kind of inertia—that characterized the AAAS in the late nineteenth century. As permanent secretary for over twenty-five years, he sustained the *Proceedings* and the annual meetings as the core of the association, and he facilitated the now intermittent public policy commentary that had characterized its early years. Most of the time he also struggled to maintain its financial base and enlarge its role in promoting a research agenda for science. Putnam studied at Harvard under Louis Agassiz and attributed his lifelong, pioneering interest in anthropology to events that occurred during his attendance at a AAAS meeting in Montreal in 1857. Together with several other former students of Agassiz, Putnam founded *American Naturalist*, working during his early career with the Museum of the Essex Institute in Salem and later adding responsibilities as curator of the Peabody Museum of American Archaeology and Ethnology in Cambridge.[78] At the AAAS meeting in Salem in 1869, the evidently able young editor and administrator was asked to substitute for long-time permanent secretary Joseph Lovering, then traveling in Europe. In the next few years, he also assisted at annual meetings.

In 1872, recognizing Putnam's growing role, the AAAS elected him permanent secretary for the following year. He held the post for a generation, until he resigned in 1897 and was elected president for 1898. Already an experienced editor, he assumed responsibility for publishing the AAAS *Proceedings* through the Salem Press, Printing, and Publishing Company, which he had organized; no one seemed concerned about the potential conflict of interest. He also arranged for the Peabody Academy of Sciences to hold the official records—records that have since disappeared and make the internal history of the AAAS in the nineteenth century difficult to trace in detail beyond the published record.[79] Putnam was by all accounts a cordial and extraordinarily hardworking anthropologist and editor, and his commitment to the organization that some years could not even pay his small promised salary voted in 1874 (and which he apparently used to hire an assistant, Miss C. A. Watson, who then served for twenty-one years) is unquestionable.[80] In 1875, a Detroit *Evening News* reporter described him as a "wiry, nervous, black-haired, black-eyed, intense looking little fellow—he bears somewhat the same relation to the Association as a brisk, strong little tug does to a fleet of heavy barges. They are all heavily loaded with valuable freight, but bless you, how slow they would be getting around if he wasn't there—he is as energetic and industrious in his special lines of study as he is as Secretary of the Association."[81] In later years, however, as his multiple positions required more of his time and attention, Putnam became less attentive to the AAAS.

Nonetheless, it did indeed fall to Putnam to arrange for speakers, to negotiate presidential nominations, to handle all official correspondence, and to manage the membership records and budget of the association. All this he did while maintaining a myriad of institutional responsibilities relating to his management of the Salem Press and the Essex Institute in Salem, the Peabody Museum in Cambridge, and a continuously growing list of curatorial and leadership roles in anthropology and archaeology—activities that gained him election to the National Academy of Sciences in 1885. Putnam was himself an indicator of the changing nature of the sciences, firmly positioned in a natural history tradition and trained by the famous Louis Agassiz, but also viewed as one of the founders of an entirely new discipline, anthropology. It was perhaps fortunate that the AAAS found someone with his temperament and intellectual predilections, someone able to operate successfully as an organizational liaison in the latter decades of the century. His term coincided with the remarkable and often turbulent expansion in American science that left a myriad of specialized societies as its legacy, societies that would directly and indirectly challenge the AAAS.

Scientific Sectionalism

Perhaps the most intractable problem facing the AAAS was based in circumstances well beyond its influence, namely the rapidly increasing specialization of science. The National Academy of Sciences, with meetings behind closed doors, at most twice a year, and poorly attended during the nineteenth century, proved not to be the chief competitor for the energy and loyalty of most AAAS members that some had initially feared. Throughout the century, the American Philosophical Society and various natural history societies, some with national aspirations and others avowedly local in membership, nurtured scientific activity. After midcentury, a new group of state academies of science became important regional forces, especially in the South and Midwest.[82] Most of these regional groups, however, benefited from the moveable feast of the AAAS, and thus worked with it to build cordial connections between state and national organizations with overlapping agendas for research, education, and popularization.

The 1887 meeting in New York provides a good demonstration of the indirect influence exerted by the AAAS, an effect that went beyond its own specific actions. Surprisingly, this was the first time the AAAS had met in the metropolis. A powerful consortium of scientific leaders coordinated plans through the old but newly revived New York Academy of Sciences. John Strong Newberry, the Academy president and a former AAAS president, had a particular goal in the invitation, intending the visiting AAAS to help "popularize science by making the public aware of the number and intellectual weight of the local scientists." The successful AAAS visit in fact gave strength to a movement to create a Scientific Alliance (federation) of New York area scientific institutions that has persisted in local leadership down to the present day.[83] Indeed, the meeting was an impressive demonstration of scientific accomplishment, including a presentation by physicist Albert A. Michelson and chemist Edward W. Morley of the results of their experiments on light waves, conducted just one month earlier. Notable visiting scientists praised the local universities, botanical gardens, museums, and societies that they toured, and an enthusiastic public reciprocated. For local institutions, it seems, and especially those that shared a commitment to public science and even relied on local patrons, the AAAS offered an unmatched imprimatur. Both national and local scientific societies benefited when the AAAS came to town.

What did begin to pull at the very center of the AAAS was specialization. Chemists, geologists, entomologists, and others began to identify themselves more with their individual disciplines than with the eclectic base of "science." The generation that had acquired Ph.D. degrees at European or the new American research universities was, in turn, finding work in highly focused federal scientific bureaus and in the new disciplinary departments that came to characterize not only transformed private colleges but also the state universities. The pool of specialized scientists increased rapidly in the last two decades of the century, and this group was less patient with the mode of operation within the AAAS. Communication needed to be efficient, and thus meetings with colleagues whose research and professional interests were similar reinforced a movement toward highly specialized societies. As these societies became active and grew, attendance at the AAAS meetings slipped into decline.

From its founding, the AAAS had recognized broad distinctions among the scientists. The original constitution had established two sections, namely "Natural History, Geology, &c." and "General Physics &c." These broad categories made room for a wide range of subjects from statistics and ethnology to linguistics and anthropology, as well as the more common categories of geology, zoology, botany, physics, astronomy, and chemistry. By 1860 two distinguishable tracks were in place: "Section A. Mathematics, Physics, and Chemistry" with subsections for 1) Mathematics and Physics, 2) Physics of the Globe, and 3) Chemistry; and "Section B. Natural History" with subsections for 1) Geology and Paleontology, 2) Physical Geography, and 3) Zoology and Physiology. "Section C. Practical Mechanics" had also been added as an incentive for engineering members to join.

The issue of managing the particular scientific interests of members had come to the forefront in 1873, when two essentially local societies of entomologists asked to meet with the AAAS and to form an "Entomological Club of the AAAS." As noted earlier, there were complicated motives for forming such a group related to the issue of standardization of nomenclature, and in this instance certain entomologists sought to work through the nationally established AAAS in resolving the scientific principles and establishing procedures for their field. The example of working side by side with the AAAS, using its committee structure, but also maintaining some independent standing for discussions among specialists at each annual meeting, seemed to provide a middle ground that satisfied the entomologists. Then, on August 1, 1874, at Northumberland, Pennsylavania, a group of chemists, celebrating the centennial of Joseph Priestley's discovery of oxygen, discussed the possibility of establishing a national association, a plan that culminated in the American Chemical Society in 1876. This organization remained largely New York-based until the 1890s when several AAAS members from Section C reorganized the American Chemical Society, strengthening the newer society but diverting their energy and iniative away from AAAS activities The forces for specialization were growing, strengthening those societies recently established and creating new ones. Frederic Ward Putnam worked to maintain a space for special scientific interests, helping his own specialty group of anthropology become first an active subfield within the AAAS in the late 1870s and then a scientific section. Under his administration, the AAAS did not resist the movement toward specialization and sometimes even willingly served as incubator for emerging fields—but leaders also fretted as older disciplines spread their wings and tested their institutional independence and endurance beyond its auspices.

In 1880, under the leadership of John L. LeConte, entomology requested its own section, and the ensuing discussion led to a broad affirmation that the AAAS "open the doors still wider, and, by the addition of other departments of research, give a still broader scope to its objects." Following that discussion, in 1881, the geologists also commented publicly on their need for a distinct group, while stressing reliance on the "elder brother" association. The commentary was ambiguous: "We geologists cannot get along without you chemists, you biologists, you physicists, and you entomologists. We need you all. But we had [other] objects in view." Indeed they did, and within a decade the Geological Society of America was founded (1888), although it continued to hold meetings in conjunction with the AAAS. Following the 1881 discussion about the aspirations of specialists, the AAAS members readily voted into place a constitutional change that created nine sections and increased the size of the standing committee. Giving more visibility to the sectional meetings, they voted that the chair of each section (who also served as vice president of the overall association) should give an annual address of general interest. The nine sections recommended in 1881 (actually echoing an experimental arrangement from the early 1850s) and put in place for the

1882 meeting were:

Section A   Mathematics and Astronomy

Section B   Physics

Section C   Chemistry

Section D   Mechanical Science

Section E   Geology and Geography

Section F   Biology

Section G   Microscopy and Histology

Section H   Anthropology

Section I   Economic Science and Statistics

Each section could, and most did, create additional subsections. Even the section categories would be continually renegotiated. In 1885 engineering was added to mechanical science in an effort to attract that growing occupational group, and microscopy and histology were linked with biology. By then the botanists had formed the American Botanical Club, which evolved into the American Botanical Society in 1892, and the following year, the AAAS leaders divided biology into sections for Zoology and Botany. The standing committee became the council in 1887. [84]

During the 1880s, too, other organizations began to meet in conjunction with the AAAS. The widespread Agassiz Clubs, primarily intended to attract young people to the natural sciences, held their annual convention the day before AAAS meetings began.[85] Other new groups such as the Society for the Promotion of Agricultural Science held their annual meeting at an overlapping time and place. Like the semiformal "clubs" formed by the entomologists and botanists, other groups afforded their members a chance to participate in both meetings.

Such moves and collaborations were not sufficient to satisfy everyone within the AAAS. In 1883, a small group formed what would become the American Society of Naturalists. The motivation of this group, with a significant proportion of college teachers, some at emerging state and research universities, had less to do with specialization than with the level of AAAS activity. While the category of "Fellow" had been intended to distance amateur from professional, the large numbers who had subsequently entered that category showed that it did not distinguish very well between mediocre and very good—at least not the way some had hoped it might. As one of the ASN leaders made clear, the new group was to be distinguished from the AAAS, which "comprises a very large and varied membership" that made it difficult to "pursue any definite line of research without interruption within that body." Another issue may well have been that the AAAS meetings were held in late summer, a time when some conducted field research, planned family vacations, or prepared for the coming academic year. By contrast, the American Society of Naturalists usually met during the week between Christmas and New Year's Day, and by the 1890s, it had established a time when several of the new specialized societies could join with it as affiliated societies and benefit from the arrangements made by the umbrella society. The new organization became what Toby Appel has designated a "half-way house" whose very existence encouraged the formation of other disciplinary societies, at least in biologically related fields.[86]

Within the AAAS, the mechanism for connecting the newly generated organizations remained under discussion. Leaders kept up a brave front, convinced that the AAAS remained valuable to scientists and important as well for the dissemination of science. As one argued in 1888, "there is strength in union; and while any small portion might flourish for a time as an organization of specialists, entirely independent of the larger body of workers and supporters, it would be a very short-sighted policy for any such body to cut the strings by which their own special work is bound to the still larger and greater work of the general advancement of human knowledge."[87] Nonetheless, independent scientific groups formed in virtually every discipline, and while some groups were short-lived, many would become part of the basic profile of twentieth-century scientific institutional life. The AAAS, created to help scientists expand beyond regional and intellectual constraints, now found it difficult to sustain the multidisciplinary discussions and specialized needs of the scientific community; discussions of interdisciplinarity would wait until the next century.

The AAAS held the middle ground through the nineteenth century. Resisting the pull away from open membership and amateur participation, it nonetheless positioned itself to consult on the issues of standardization and patronage that most concerned professionals. In the debates over religion, leaders gave latitude to speakers to express their opinions from atheism to deeply held Christian beliefs, but authorized publication only of scientific papers and plenary addresses that would not embarrass or discredit the association. Always the association emphasized the role of expertise. When asked for a resolution on vivisection, for example, the members went on record as deploring the "cruelty and needless vivisection experiments in the public schools," but asked Congress to leave it to "scientific men to decide" appropriate use of the technique for research rather than passing restrictive legislation. The AAAS council sustained conservation efforts but was never in the vanguard of activism. As might be expected in a broad-based society, the AAAS membership typically reflected a mainstream position, working in the range of noncontroversial decisions while trying to advance the shared interests of members who bridged a range of disciplines and were located in a variety of institutional and geographic settings. Even its governance structure emphasized the operating principles and practices that, at least nominally, involved a broad, democratic base. In the 1890s, the unwieldy AAAS council included all living past presidents, its five officers, and four representatives from each of the specialized sections, and all its decisions were ratified by an open general session.

Frederic Ward Putnam, perhaps inspired by his experiences with New England town meetings, was committed to the direct democracy put forward by the Rogers brothers and others and seems to have genuinely meant his offer to handle each individual member's concerns. Trying to run the association single-handedly with but one staff member, Putnam in the 1890s nonetheless also took charge of ethnology for the Chicago World's Columbian Exposition of 1893 and accepted curatorial responsibilities for anthropology at New York's American Museum of Natural History, in addition to his responsibilities in Salem and Cambridge. It may have been his regional bias or simply lack of initiative that kept the AAAS meetings largely in the North and East, with only three meetings just beyond the Mississippi River (in St. Louis, Dubuque, and Minneapolis) and one in the South (Nashville).

In the 1890s, despite nearly record numbers at the Washington, D.C., meeting in 1891 and the following year in Rochester, New York, the AAAS was in trouble. Reflecting on fifty years of history since the founding of the AAGN, Thomas C. Mendenhall's presidential address expressed a fundamental dissatisfaction over the ability of the association to meet its original intention "to procure for the labors of scientific men increased facilities and a wider usefulness."[88] Centrifugal forces seemed to be pulling apart the efforts to maintain cross-disciplinary scientific communication over the divides created by increasingly distinctive disciplines, rapidly realigning research and educational institutions, and the continuing geographic obstacles of getting scientists together.[89] As economic depression and panic took hold in 1893, the attendance at annual AAAS meetings fell, the *Proceedings* appeared later and later, and some younger reform-minded members determined to bring about some change. The unconventional and ambitious W J (he used only initials without periods) McGee of the Bureau of American Ethnology was bothered by specialization and the growing isolation of the specialist, warning that expertise was "soon blasted by the poison of its own egoism, unless the richer part of substance is guided toward the general mass of society."[90] It was time, he and his cohort believed, to strengthen institutions that would keep lines of communication open, create a unified voice for science, and establish working relationships among the increasingly well-established specialities.

Science magazine, which had struggled financially for a number of years and had been sold in 1894 to James McKeen Cattell, was provided a subsidy by the AAAS, which in its turn gained another forum for its activities. For the committee that recommended this new policy, chaired by McGee, the mutual advantages seemed quite clear. The editorial and business control shifted to the AAAS. Moreover, the new editor, Cattell, brought dynamic energy and an assertive advocacy style that spilled over to the association itself. In a *Science* editorial, he argued that the AAAS had filled a unique function and was distinct from the National Academy of Sciences because it was independent, not at the beck and call of government. He made a virtue of its disciplinary and geographically diverse membership, concluding that "in view of the rapidly increasing tendency toward centralization of scientific work in the government . . . it will be well worth while to maintain the Association. . . . " Moreover, the challenge by the American Society of Naturalists to create some alternative umbrella organization, with which various biological societies might affiliate, also failed to fully accomplish its goal. As the following chapter demonstrates, when the AAAS established a multisociety Convocation Week in 1901, it reasserted its commitment to interdisciplinary programs and promoted, at least temporarily, cooperation and affiliation among the newcomers and the venerable parliament of science. [91]

Putnam, perhaps inspired by the challenges and new directions opened by cooperation with *Science*, wrote an essay for the magazine in 1895 outlining the history, goals, and accomplishments of the AAAS over nearly fifty years.[92] He also tried to bring higher visibility by persuading the eminent chemist Wolcott Gibbs to run for president, pointing out that Gibbs's leadership would help to "renew interest of the leaders of American science in the Association" at a time when the AAAS was being squeezed between the Academy and the new specialized societies.[93] Gibbs, one of the few members from the 1850s still alive and active, had been elected president of the NAS the previous year and agreed to serve, making him the first to hold both leadership positions concurrently. Then, however, Gibbs failed to attend the meeting over which he was supposed to preside, perhaps an expression of displeasure over a contested decision to hold that meeting in Detroit; the following year, he delivered his presidential address on a technical aspect of his research.

Putnam marked time until his twenty-fifth year, having attended every meeting and edited every volume during his term, and then stepped down from the position of permanent secretary. In 1898 Putnam introduced his successor as permanent secretary, Leland O. Howard, a well-respected entomologist who had already held various offices in the AAAS section on zoology. Well connected through his active participation in the Cosmos Club and just a few years earlier made head of the Division of Entomology in the Department of Agriculture because of his work on insect predation, Howard seemed a promising choice to provide new, active leadership.[94] It was a bittersweet moment as the association returned to Boston and celebrated its fiftieth anniversary in the city where the active AAGN had resolved itself into the AAAS, thus commemorating a significant record of achievement but with falling membership and attendance signaling an uncertain future.

Conclusion

During the course of the nineteenth century, the AAAS was, as some envisioned it, a "parliament of science" wherein broad and sometimes conflicting points of view might be debated and resolutions found. Its resolutions and committee reports reflected the commitment but also the ambivalence within the scientific community, particularly toward issues of definition of science, status among scientists, popularization, and advocacy about the role of the federal government in support of science. It had held the scientific center—but that position was maintained in constant tension between the ambitious tasks of facilitating scientific work and of providing public access to scientific results. Viewed one way, the AAAS advanced science by offering scholarly and technical presentations that became less and less accessible to the local press and public, opting to articulate and demonstrate high quality through a tightly controlled operation of presentations and publications that insured its reputation. Viewed from a different perspective, the AAAS advanced science by maintaining an open-door practice where an often eclectic mix of papers allowed less than well-connected scientists to have an audience and debate some of the most important issues of the day, including evolution. The result was that the organization suffered from the dissatisfactions of the most outspoken proponents of popularization and of elite science, even as it held its moderate membership. While it remained singularly visible as a forum for science, the AAAS had not acquired the scientific authority envisioned by its most ambitious founders.

Perhaps inevitably, given institutional patterns abroad as well as in the United States, many scientists put their best energies into creating the new, more specialized societies—and without a charismatic spokesperson, the benefits of the AAAS were not often strongly articulated in the latter part of the nineteenth century. The result was a somewhat restrained association working to weather the turbulent and uncertain conditions for science in the period before progressive social and political incentives would position scientific expertise more prominently. By the early twentieth century, the reforming leaders were concerned about the effects of specialization, seeking ways to advance discussion across the new boundaries, and finding new ways to increase public support for science.

With its regularly published journal, its genuinely national membership base, and its long tradition of speaking for a broad spectrum of scientists, the resilient AAAS seemed uniquely positioned to address the concerns of scientists and their many public audiences after the turn of the century. Such a forecast, however, would be premature. Affiliation with *Science* proved a dramatic turning point that brought more stable funding and regular coverage for the AAAS, but with it also came the ambitious young James McKeen Cattell, whose long editorship would sometimes compromise the elected leadership within the association.

CHAPTER 2

Promoting Science in a New Century

THE MIDDLE YEARS OF THE AAAS

MICHAEL M. SOKAL

$\mathcal{A}$s the twentieth century opened, the American scientific community had firmly established its professional (if not intellectual) independence from Europe. Scientists at the new private research-based universities like Johns Hopkins, Clark, Chicago, and Stanford had shown that Americans could initiate, conduct, and complete their own research programs. The growth of the land-grant universities established during the Civil War had opened up higher education to the rapidly expanding middle class and had also demonstrated the value of research-based knowledge for American industrial, agricultural, and social development. Federal agencies like the U.S. Geological Survey and the Division (later Bureau) of Forestry had proved effective at combining basic research and resource management, thus laying the foundation for the Progressive commitment to rational government oversight that soon guided the early years of the century.

The sciences in America had also started to move beyond their European roots. The Michelson-Morley experiment had begun, however inadvertently, to challenge older conceptions of the ether, and American biologists working both in "pure science" research laboratories and at state-supported agricultural experiment stations had moved the revolution started by Darwin to its next stage through their studies of heredity. And though the transformation of the study of the human mind into research on human and animal behavior lay a decade or so in the future, by 1900, over a dozen universities were sponsoring productive psychological laboratories.

Of particular importance to the American Association for the Advancement of Science was the near-universal professionalization of science. To do cutting-edge research, individual scientists could no longer work as their mid-nineteenth-century predecessors had, stealing time from careers in law and the ministry. Such amateurs had been displaced by scientists working at universities, teaching colleges, natural history museums, government research agencies, and the new industrial research laboratories of such technology-based firms as General Electric, American Telephone and Telegraph, and several chemical and pharmaceutical companies. With professionalization came specialization, and by the close of the nineteenth century, specialized scientific societies in (among other areas) chemistry, physics, and the several biological disciplines had all begun to separate themselves from the AAAS. Their members preferred to present their latest research results before their professional colleagues, and thus called into question the AAAS's traditional function, as embodied in its initial objectives.

Adapting for Science and the Scientific Community

As the century began, AAAS leaders thus realized that if the association were to survive, it had to respond to this challenge. Many, such as anthropologist W J McGee, thus sought to help it transform itself into an organization to provide a forum for all science, offering a stable interdisciplinary (or multidisciplinary) center against the splintering factionalism of increasingly narrow disciplinary interests.[1] And as concerns of previous generations of AAAS leaders grew less significant, other challenges also emerged. For example, by 1900, earlier tensions had declined between those members who emphasized the AAAS's "democratic" outreach mission and those who stressed its service to a more limited active research community. But by that time the AAAS had to confront the problem of maintaining its large membership and broad representation of the scientific community in the face of the resurgence of a scientific elite that believed it knew best how to advance science by directing research and controlling the resources required by increasingly large-scale science. In the same way, even as the need to define science itself—so important to the AAAS's founders—declined, early-twentieth-century AAAS members found themselves having to promote the value of scientific approaches in such practical areas as food safety and land management.

In the face of these challenges, by 1898, the American Association for the Advancement of Science and its leaders had begun to respond to the late-nineteenth-century changes in its institutional environment and to introduce reforms that allowed it to compete more effectively with the specialized societies that emerged during the previous two decades.

The AAAS and Science as Advocates for Science

In many ways, the association's most significant adaptation during this period involved the development of its symbiotic relationship with *Science*, the nation's leading general scientific periodical. Through their close ties and interlocking leadership, the AAAS and the journal soon became better able to work for the growth of the American scientific community and to promote its members' interests, thus extending beyond its founders' hopes the Association's influence on the course and conduct of American science.

Even as American scientists began to identify themselves with their particular disciplines, many called for a general periodical to serve (and help build) the entire rapidly growing scientific community, much as the AAAS reformers saw the association serving as an "umbrella" for all sciences and scientists. But defining the niche to be filled proved difficult. During the 1870s and 1880s, at least seven journals—all privately owned, like most American scientific periodicals of the period—tried and failed to bring the sciences together. From 1880, *Science*'s successive editors and owners—including Thomas A. Edison and Alexander Graham Bell—all hoped to respond to their colleagues' interests. By 1885, however, after Bell and his partners had lost at least $40,000 on *Science*, they allowed scientific amateur N.D.C. Hodges to take over the journal. But Hodges lacked professional credentials and knew little of the American scientific community, and the Panic of 1893 reduced his income sharply. He discontinued *Science* in March 1894 and soon thereafter issued a circular headed "'Science' The Journal a Gift to the Scientific World," whose equally awkward text sought to explain away his failure and win additional support.[2] Many scientists regretted that *Science* had followed its predecessors into failure, but few produced concrete suggestions for action.[3]

One group that did respond, however, was "a committee of founders and friends of the journal *Science*" that met at the August 1894 AAAS meeting to "secur[e its] continuation." This group's resolutions called for the association to pay an annual subsidy to *Science* and for the journal to publish AAAS papers and reports, and recommended the appointment both of associate editors from each AAAS section and of an "executive committee" to oversee business and editorial policy.[4] The association endorsed these resolutions, and late that month Hodges made one last attempt to win support for his editorship under these terms. But American scientists had lost faith in him. His efforts failed, and that fall he sold the journal for five hundred dollars to James McKeen Cattell. For the next fifty years Cattell dominated the AAAS in ways that at first helped ensure the association's survival and promoted its interests and those of the American scientific community, but later threatened its existence. In many ways, Cattell became the most significant figure in AAAS history, and through his editorship of *Science* the AAAS did much to advance the cause of science in America.[5]

In 1894, not many American scientists knew Cattell, then professor of experimental psychology at Columbia University. Columbia geophysicist (and newly elected AAAS treasurer) Robert S. Woodward remembered years later the "strange but attractive young man, who came into my office . . . in the autumn of 1894 like a meteor out of a clear sky, and astonished me by inquiring whether he might not undertake the resuscitation of the defunct journal *Science*."[6] But they soon grew familiar with his self-assured (and at times self-righteous) manner and, especially, his willingness to work long and hard for *Science* and for the association.[7] He had come to Columbia in 1891, helped found the American Psychological Association a year later, edited the APA's first *Proceedings*, and with Princeton's James Mark Baldwin founded the *Psychological Review* in 1894.[8] Several months later, as *Science* ceased publication, Cattell offered to take it on without salary if "(1) there should be no actual money loss and (2) leading men of science would consent to contribute."[9] As soon as he bought *Science*, the thirty-four-year-old Cattell began to devote to the journal a new combination of energy and professional status, and that fall he won support from "leading men of science." He emphasized the "umbrella" function of *Science* as he claimed both that "the lack of a journal such as *Nature* in America would seriously interfere with the harmonious development of science as a whole," and that "when there are so many special journals we need all the more a journal which will report on the progress of science as a whole."[10]

By December, Cattell had recruited a larger and much more diverse and distinguished editorial committee than the AAAS committee had proposed; most of the eminent scholars he gathered to "represent" their sciences worked either at the federal bureaus or the new research universities. Under his editorship, *Science* served, as none of its predecessors had, to advance science in America by publishing research results. More important, it also advanced the cause of science by establishing a valuable sense of community among American scientists. For example, from Cattell's first issue "Scientific Notes and News" proved to be *Science*'s most interesting section. For many years it covered the comings and goings of American scientists, news that meant much to many when the United States had fewer than five thousand scientists, the AAAS itself had fewer than two thousand members, and many of these men and women knew each other. Previous editors had published similar notes, but Cattell greatly expanded their coverage. They reported scientific appointments, the progress of scientific exploring expeditions, and gifts and bequests to scientific institutions. For federal scientists, *Science* published notices of Civil Service examinations for scientific positions, and a subsection headed "Educational News" provided analogous coverage. At times these notes also included editorial comments. But news items—reprinted from the daily press, from *Nature,* and from other journals, or gathered through Cattell's extensive reading and correspondence with his editorial committee and others—always dominated this section.[11] As early as 1897, as the new editors of the *American Naturalist* tried to define a niche for their journal, they readily admitted that the niche of "the weekly scientific newspaper . . . is already admirably supplied in this country."[12] All who read this comment knew that it referred to Cattell's *Science.*

Cattell also edited *Science* "to report on the progress of science for men of science." As he told a colleague, "You cannot expect to learn about [your specialty] from such a journal. But you can tell me and others working in related departments what we ought to know."[13] Through the twentieth century's first decade, *Science* thus published several series of "Current Notes" on various topics, including anthropology, physiography, psychology, and evolutionary theory. These served much as did the "Research News" sections commissioned by later *Science* editors. *Science's* "Discussion and Correspondence" columns also attracted many contributions and a large readership, as Cattell actively encouraged debate, at times by sending proofs to authors of books reviewed "in case [they] wished to discuss some of the points."[14] Like almost all scientific journals of the period, Cattell's *Science* did not employ peer review. Most editors, including Cattell, instead had to stimulate others to submit material, and he published just about any item vouched for by members of his editorial committee or sent in by scientists employed by federal agencies or reputable universities. This policy had its problems, and at times readers complained about specific items. To many, *Science's* book review section seemed notably weak, and authors and publishers—especially those who advertised their books in *Science*—sometimes complained about the journal's sloppy and often unthoughtful reviews. But Cattell had to fill *Science's* pages, and even as he admitted to a colleague that "there are many things . . . in

Science of which I do not approve," he concluded that "an editor must be an opportunist."[15]

Although *Science* always dominated his attention, Cattell knew that a strong and broad-based AAAS would enhance the journal's financial and scientific success. He thus worked from the start to intertwine *Science*'s interests with those of the AAAS, and arranged to receive the AAAS subsidy recommended by the "committee of founders and friends," even as he made sure that it would not "encumber" his ownership. *Science* also continued to publish AAAS annual-meeting material, including the retiring president's address, the addresses of the vice president (section chair) for each section, and abstracts of papers presented. Through the late 1890s, Cattell wrote a series of editorials supporting efforts to reform the association and, especially, its governance. In discussing this topic, Cattell, like Frederic Ward Putnam (who had been permanent secretary for twenty-five years), used the rhetoric of democracy that AAAS leaders had long employed. As he wrote, "the Association should be a true democracy." But for Cattell this phrase meant "representative democracy," and when AAAS members voted to reject one of the reformers' recommendations, his editorial argued that the membership, having chosen its delegates, "their deliberative action should not be reversed by inconsiderate impulse. As Huxley has said, 'there may be wisdom in a multitude of counselors, but it is usually in one or two of them.' Folly is also more likely to be concentrated in a crowd, and unfortunately folly is more contagious than wisdom."[16]

Cattell continually cited the association's commitment to democracy through the middle of the twentieth century, even as he always believed that its members should defer to its leaders. This tension in the meaning of "democracy" had a series of major impacts on the AAAS's evolution.

By 1896, *Science* under Cattell was publishing more AAAS material than it ever had, and late that year, the association voted to issue only condensed versions of the annual AAAS *Proceedings*. The reformers also began considering how the AAAS might extend its arrangement with *Science*. One proposal called for the association to send to its members annually the eight issues of *Science* containing most AAAS material, to pay Cattell fifty cents per member for these issues, and to offer full-year subscriptions for $4.50. The plan met stiff opposition, however, and in 1897, AAAS members not only rejected it but also voted to restore the *Proceedings* "to the former method of publication."[17] The reformers thus delayed further effort to extend the association's relationship with *Science* until 1900, when "the future of the

Association and of *Science* were definitely assured." Cattell then formally proposed "that *Science* should be the official organ of the Association . . . and be sent to all members at a rate that would cover the actual cost," estimated at $2.50 annually, rather than the annual subscription rate of five dollars. He also spelled out the benefits of such an arrangement for "science in America, the Journal, and the Association." To many the advantages to the first were obvious, and he noted that "while the plan does not propose any money profit to the Journal from subscriptions," nonetheless "some profit would follow in the advertising department." He thus emphasized how "the Association would gain greatly by the plan" and avoid "break[ing] up into separate societies." He noted that "the weekly receipt of *Science* would give those who do not regularly attend the meetings [that is, most AAAS members] a definite and valuable return for their membership," stressed the many "men of science who subscribe to the journal, who would doubtless join the Association if the Arrangement was made," and concluded that "I see no reason why membership in the Association should not be enlarged very greatly." He assumed that AAAS annual dues would increase from $3.00 to $5.00 to cover the new costs, but noted that the association would have no editorial and publishing expenses and that annual dues for the British Association for the Advancement of Science (which did not provide a subscription to a journal) were also about $5.00. He claimed that "membership in the Association worth $5.00 and a Journal worth $5, both secured with one $5.00 fee, would prove attractive."[18]

At the association's 1900 meeting, the Council announced both that it would "furnish the journal *Science* to all members . . . at the rate of $2 per year each" without any dues increase, and that *Science* would henceforth carry the subtitle "publishing the official notices and proceedings of the AAAS." Permanent Secretary Howard noted that AAAS members "unable to attend meetings" often "wish[ed] the advantages of membership were more tangible" and expected that "the weekly receipt of *Science* . . . will silence such criticism."[19] This arrangement involved serious risks for both the association and the editor, as it meant that the AAAS would receive only $1.00 (rather than $3.00) annually from each member, and that Cattell would receive only $2.00 (rather than $5.00) for each annual subscription to *Science*. All knew that only great membership growth would lead to its ultimate success, and within a year such growth surpassed both Cattell's and Howard's hopes. During 1901, AAAS membership almost doubled, and it continued to grow rapidly thereafter. It reached 4,000 in 1903, 5,000 in 1906—when the first edition of *American Men of Science* (which Cattell also edited) included only about

4,000 names—and 6,000 in 1909. The association's Treasurer's Balance (which acted as an endowment) also grew from about $5,400 in 1896 to more than $16,000 in 1906, as did AAAS spending on other activities.

But despite the naming of a "Committee on the Relations of the Journal, *Science*, with the Association," the AAAS never had "strict supervision of [its] conduct," and Cattell always retained his editorial independence. To be sure, in 1904, at the AAAS committee's recommendation, he added to its Editorial Committee the chairs of all AAAS sections, a group (so he claimed) that comprised "the most efficient and active men of science in the country." And though he noted that these additions insured "that all branches of science and all parts of the country [were] adequately represented," he still rarely consulted this committee formally. Soon thereafter, he even reported that "*Science* is now so well established as the representative organ of American men of science that it seems unnecessary to print each week the names of the editorial committee and the responsible editor."[20] As these developments suggest, during the first years of Cattell's editorship, *Science* had gradually assumed a central role in the American scientific community. In 1908, however, as it strove to increase the public understanding of science, the association agreed that its members could request, in place of *Science*, *The Popular Science Monthly*, which Cattell had owned and edited since 1900. The *Monthly* published more advertising than did *Science*, and increasing its circulation allowed Cattell to raise its advertising rates.[21] Though the association had no formal links to the *Monthly*, the arrangement served both well for years.

Promoting Science and the Association Through Convocation Week

As *Science* grew in importance, AAAS reformers looked to Permanent Secretary L. O. Howard to help revitalize other association programs. Like Putnam in his last years in office, however, Howard never devoted his full attention to AAAS affairs. He continued his well-respected career in entomology, and at the Department of Agriculture helped develop important new methods of pest control. In 1894, he had become chief of the department's Division of Entomology, and in 1904, he oversaw its transformation into a formal bureau. Indeed, he was so taken with the bureaucratic maneuvering all too common among federal scientific agencies that he did not retire as chief until 1927. He also greatly enjoyed his active social life at Washington's Cosmos Club, and AAAS interests thus always remained for him a secondary or even tertiary interest.[22] Reinvigorating the association's long-standing primary program, its annual meetings, and adapting them to the early twentieth century's new environment, thus fell to other AAAS leaders, including Cattell and, especially, Harvard embryologist Charles S. Minot. In particular, as concerns grew about these meetings' relevance for the growing scientific community, *Science* editorials through the period compared them to those of the American Society of Naturalists, which "represent an attempt to keep a group of sciences in mutually helpful relations . . . [and] men of science in different but related departments . . . into personal contact." The editorials also noted the problems that late-August meetings caused university scientists and, in discussing alternative times, emphasized how late December meetings would mesh readily with most academic schedules.[23]

Late in 1900, the AAAS formed a committee to develop a specific proposal, and its chairman, Minot, soon began using the phrase "Convocation Week" for "the week in which New Years Day falls."[24] The committee proposed that all national scientific societies meet concurrently with the AAAS during that week and that all American universities adjust their calendars to facilitate such meetings. By 1902, most major universities had. But even earlier, in August 1901, the association announced plans to hold its first winter meeting in Washington at the end of 1902. New constitutional amendments also gave affiliated societies representation on the AAAS Council—thus formally opening and enlarging its "umbrella"—and many American scientists and most scientific organizations welcomed these changes. In December 1902, 989 AAAS members and fellows registered for the association's first winter meeting, and twenty-four affiliated societies (with 363 additional registrants) held concurrent meetings. *Nature* called it the association's "most successful meeting ever" and praised the "many notable papers" presented and the "character of the proceedings." President Theodore Roosevelt received AAAS members at the White House, and one AAAS officer claimed that the "great majority of the working scientific men of the country" gathered for the occasion. And though not all affiliated societies always (or in some cases often) met with the AAAS, and though attendance did fall off at later meetings in cities like New Orleans, late-December meetings worked well for the AAAS and for many of its affiliated societies for many years. The joint meetings continued into the 1970s, providing a forum that helped maintain a stable center of the scientific community against the centrifugal forces of specialization.[25]

But the success of Convocation Week had its costs. Some members complained about being away from their families during the Christmas holidays and, in choosing to respond to the interests of academic scientists, the association clearly slighted those employed by industry and government agencies. To be sure, through the early 1900s universities annually employed more and more scientists, and this sheer growth led to their growing dominance in

AAAS affairs. As the move toward Convocation Week reinforced this trend, tensions between government and academic scientists remained latent, and for many years industrial scientists rarely played significant roles in AAAS affairs.

In the meantime, some sciences relied more on Convocation Week meetings than others, and the dependence of each varied through time. At times this dependence remained significant despite the views of a particular science's influentials. Even earlier, for example, despite the criticisms of the AAAS by leading physical scientists in the 1870s and 1880s, Albert A. Michelson and Edward W. Morley presented the results of their epoch-making ether-drift experiments at the 1887 AAAS meeting. In 1888, Michelson used his AAAS vice-presidential address to issue "A Plea for Light Waves." Four years later, astronomer George Ellery Hale impressed colleagues from across the country with his spectroheliograph, and Michelson, for one, at once saw its potential to revolutionize astrophysics.[26]

In the new century, at the 1908 joint AAAS–American Physical Society meeting, Gilbert N. Lewis and Richard C. Tolman first introduced American audiences to Albert Einstein's special theory of relativity. And as some American physicists became convinced of Einstein's claims and others denounced them in the name of "common sense," in December 1911 the APS and the AAAS section on physics sponsored a joint symposium on "The Ether." The session featured papers by Michelson, Morley, Lewis, and two young recent converts to relativity, Daniel F. Comstock and William S. Franklin. *Science*'s report on the symposium concluded that "some difficulty was experienced in finding common ground" among the speakers. William F. Magie, that year's section chair, had no doubts about the falseness of relativity, however, and denounced Einstein's ideas in his vice-presidential address, "The Primary Concepts of Physics." In the years that followed, however, Robert A. Millikan used AAAS-sponsored occasions to argue for less conservative views. In 1912, for example, he used his own vice-presidential address to speak on "Unitary Theories of Radiation" and to report on quantum notions that he had recently encountered in Europe. Four years later, Millikan introduced Americans to the Bohr atom through "Radiation and Atomic Structure," his American Physical Society presidential address before a joint AAAS-APS audience. But as these examples suggest, by the early 1910s and especially afterward, most AAAS section on physics sessions and symposia reflected the organizational impact of the American Physical Society, and as the century continued physicists tended to look more to APS and less to the AAAS.[27]

Biologists, on the other hand, began looking more and more to the AAAS after 1900, as the association usurped some of the functions previously played by the American Society of Naturalists. An informal "Biological Smoker" remained an important part of all AAAS meetings through the twentieth century's first half, and beginning in 1902, the naturalists themselves and the societies they spawned regularly joined the Convocation Week meetings. The members of these biological and geological societies also seemed more apt to work in relative isolation—on exploring expeditions and at faraway field stations and marine laboratories, for example—than their colleagues in the physical sciences, and they thus especially welcomed the social and intellectual excitement of Convocation Week. As these meetings typically brought together those in all biological disciplines, these researchers tended to find in AAAS meetings at least some of the kind of interdisciplinary stimulation that the AAAS reformers had praised in the 1890s. No wonder biological and geological scientists seem to have looked more to the AAAS than have physical scientists throughout its history, and to have played more active roles in its twentieth-century leadership.

Even as ever more specialized societies emerged through the century, they all met with the association, as it provided the counterweight (which many of their members sought) to the narrowness that accompanied increasing specialization. Years after the founding (at the 1892 AAAS meeting) of the Botanical Society of America, one of its leaders, John Merle Coulter, claimed that the AAAS served as "a convenient organization to bring botanists together." Deliberations among botanists at AAAS meetings led by the mid-1890s to the long-standard American Code of Botanical Nomenclature, and in the decades that followed, even as some botanists attacked this code and urged its replacement with what they informally called the "Vienna Congress Rules," they did so at AAAS meetings. Throughout this period botanical societies came and went, but they all met in conjunction with the AAAS.[28] So too did zoological groups, especially as their members often belonged to several different societies, such as the American Society of Zoologists, the American Physiological Society, the Association of American Anatomists, and (after 1931) the Genetics Society of America. Certainly in 1912 those biomedical researchers who found Convocation Week meetings less valuable than did university-based botanists and zoologists formed a Federation of American Societies for Experimental Biology as their own "umbrella." But all other efforts to sidestep the AAAS failed, largely because through these decades most members of the ASN, ASZ, and BSA still found much of professional value in the AAAS's Convocation Week meetings.[29]

In making meetings more professional, however, the AAAS seems to have downplayed its long-standing goal of bringing science to the public at large. For early-twentieth-century AAAS leaders, then, "advancement" of science meant focusing on the immediate needs of professional researchers, rather than on the place of science in broader American society. As emphasized earlier, nineteenth-century meetings had often featured public lectures on popular topics illustrated with lantern slides, open to and advertised for the residents of the host cities. Throughout the 1890s, older AAAS members continued to call on local committees to emphasize this aspect of their programs, and actively complained when supposedly popular addresses proved more suitable for scientific than for general audiences. Indeed, as early as 1874, Edward L. Youmans, a renowned popular scientific lecturer, complained that the AAAS had become nothing more than "An Association for the Promotion of Science by Original Research." Many of the reformers of the late 1890s and early 1900s seemed to share this research goal; public outreach played an increasingly smaller role in successive AAAS meetings, and the June 1900 meeting in New York dispensed with public lectures entirely. The association's move to Convocation Week meetings signaled both its inward focus and its shift toward the dynamics of scientific specialities and away from the rhythms of public discourse. This change of focus may well have been necessary, for it enabled the AAAS to adapt itself to an altered environment in which it faced serious competition. The symbiotic relationship that emerged with more specialized societies served almost all of them well for many years.[30]

Progressive-Era Politics, Governance, and Expansion

Despite this general turn inward, AAAS leaders continued to assert their traditional claim to speak for the scientific community as a whole, and especially to advocate for a role for science in the setting of national policy. As a result, as *Science* and the association's annual meetings grew ever more successful, many of the association's early-twentieth-century leaders tried to extend its general influence nationally and to advocate more forcefully for science and its practitioners. As noted earlier, through its first fifty years the AAAS had tried to shape public policy by passing resolutions calling for particular state or federal responses to issues of concern to scientists. As early as 1897, a *Science* editorial envisioned expanding the AAAS's role in the governance of and policy-making for American science through the institutionalization of "the National Academy as the 'upper house' and the American Association the 'lower house' of American science."[31] During Cattell's first years as editor, *Science* often published editorials calling for specific actions. These included the appointment of scientifically competent leaders for the Coast Survey, the Naval Observatory, the Fish Commission, and the Bureau of American Ethnology, and, most notably, for the reform of the Smithsonian's scientific bureaus. Cattell also used *Science* to attack the leaders of the Carnegie Institution of Washington for instituting policies that downplayed grants for individual researchers to emphasize support for institutions such as its Mount Wilson Solar Observatory and its Eugenics Research Station.[32] Through the twentieth century's first half decade, many American scientists seemed to assume that *Science* and its editorial committee acted for the AAAS. When pressed, however, Cattell always admitted (in print, at least) that, as editor of *Science*, he spoke only for himself. But at times he self-assuredly seemed to suggest (and, in private, actually claimed) that he acted as speaker of the "lower house" that he hoped the AAAS would become.

In 1906, the AAAS took more direct action than ever before by creating a Committee of One Hundred on National Health, which took the distinctive form of a Progressive-era pressure group. Its large membership passed resolutions calling for the federal government to devote at least as much care to the health of humans as it did (through the Department of Agriculture) to that of farm animals, and it campaigned for the creation of a National Department of Health. Though its objectives readily won wide endorsement, it ran afoul of infighting among cabinet secretaries, and by 1912 organized opposition led by the Anti-Vivisection League and the Church of Christ, Scientist, killed the proposal.[33]

In the meantime, the association's ties with *Science* and its efforts at serving the entire scientific community led to great increases in membership. Late-twentieth-century concerns about "diversity" among the association's membership remained unknown in this period, and the membership thus replicated itself as it expanded. For many years, AAAS members remained primarily white, male, and middle-class, with positions in federal scientific bureaus and, increasingly, at research universities. The association's leadership closely reflected its membership. Its governance, though officially vested in its council, remained from 1898 in the hands of the committee on policy. Despite its ostensibly democratic form, the committee's membership proved self-perpetuating, for as each member's three-year term expired, his colleagues always nominated him for reelection, and the council always ratified their choice. To be sure, not all continuing members remained constantly active. But several did, including Cattell, Minot, Herman L. Fairchild of the University of Rochester, Henry B. Ward of the University of Illinois, and Daniel T.

MacDougal of the Carnegie Institution's Desert Botanical Laboratory in Tucson. These middle-aged white men and others much like them controlled AAAS affairs at least through the 1940s.

Though other scientific societies in the 1910s and afterward actively involved women and welcomed Catholics and Jews into their leadership, the AAAS never did. The Association seems never to have consciously discriminated, and sociologist W.E.B. Du Bois—who joined the association in 1901 as perhaps its first African American member—was elected to fellowship in 1905. Women had belonged to the association at least since the 1850s, when Maria Mitchell and Margaretta Morris became members. But in 1900, when 13.4 percent of all AAAS members were fellows, only 4.2 percent of the association's 864 women members had that status. In 1915, the section on anthropology and psychology elected as its vice president Lillien J. Martin of Stanford. But though AAAS leaders seem not to have purposely restricted their focus, broader concerns for diversity did not emerge for many years, and the AAAS did not elect its first woman president until 1969.[34]

At about the same time, other signs began suggesting that the AAAS would profit from further adaptation or, at least, from stronger executive leadership. Some problems emerged even amid the association's great successes. The December 1906 meeting in New York, for example, attracted nineteen affiliated societies and perhaps 1,800 individual scientists. Howard attributed its success largely to "the desire on the part of members not only to associate with specialists in their particular department of work but to affiliate with specialists in other branches of scientific investigation." Even previous critics praised the meeting's "exhibition of recent progress in science" and, especially, the plenary session on "The Biological Significance and Control of Sex" at which representatives of all affiliated societies participated. The session itself was (according to the weekly magazine *The Independent*) "packed . . . like a subway car in rush hour." But the magazine's reporter (apparently Edwin E. Slosson, later director of the Science Service news bureau) also expressed deep dissatisfaction. Though he praised the individual speakers, he noted that each spoke "as he would in the isolation of his own lecture-room" and made "no reference to what his predecessor ha[d] said." The report thus concluded that "there was really no discussion at all" and, even more significantly, claimed that "there was apparently nothing whatever gained by bringing these men together upon the same platform and assembling the audience." Though the report went on to urge the association to encourage the productive interplay of (and, especially, debate among) speakers within a given session, this article served to question the value (or at least the form) of Convocation Week meetings, which remained (with *Science*) the association's primary function. Influential AAAS members never believed that these annual Convocation Week meetings had outlived their usefulness, and, as noted earlier, many biologists and others continued to find them extremely valuable for years. But attitudes like these apparently did much to stimulate interest in developing and enhancing significant new AAAS programs.[35]

Unfortunately, as the need for further adaptation grew, Permanent Secretary L. O. Howard devoted less and less time to AAAS affairs. University scientists had always considered him unaware of their interests, and, though membership and *Science*'s circulation continued to climb, in Cattell's view they grew all too slowly. After 1906, Cattell focused more than ever before on AAAS affairs. For example, to increase his editorial autonomy he had the AAAS discharge its long-standing "Committee on the Relations . . . with the journal *Science*." He also began telling others of Howard's failings. After serving as AAAS president for 1906, Johns Hopkins pathologist William H. Welch agreed that the AAAS needed "a systematized, central, business organization with a permanent secretary working all the time for the Association."[36]

Things improved slightly in May 1907, when Charles D. Walcott, the new Smithsonian secretary, offered space to the AAAS. Howard accepted, but rarely occupied the new office. His assistant, F. S. Hazard, shared his inadequacies, AAAS membership records (and thus *Science*'s subscription list) remained in disarray, and at times local committees for annual meetings could "not obtain any answers from Washington." Cattell claimed to "suffer more" than anyone, as AAAS members complained often "about failure to receive *Science* or answers to enquiries . . . [and] no efforts are being made to enlarge the membership." All agreed that "there are any number of ways in which the association might promote the influence of science and scientific work through proper organization," that "the usefulness and development of the association are interfered with by the inefficiency of the secretary's office," and that the association needed "a secretary who will devote his whole time to the secretary's business."[37]

Successive policy committee chairs raised concerns with Howard, but his performance never improved. In 1908, he called for payment of the next year's annual dues simply through a note in *Science*, and the reduced-rate rail fares he arranged for that year's annual meeting actually cost more, with required fees, than regular round-trip tickets. As the December 1909 annual meeting neared, Howard did conduct a membership drive in the host city of Boston. But in a throwback to the earlier era in which the AAAS first emerged, he targeted those listed in the city's Social Register, rather than the faculties of local colleges and universities.

Despite these failings, all realized that the association could not afford a full-time executive. For years, the AAAS kept its constitutionally mandated and thus hard-to-raise annual dues at $3.00, and as the American scientific community grew, its membership multiplied, despite Howard's blunders. But since 1901, the AAAS had earmarked $2.00 of each member's dues for *Science*, so even as its payments for the journal just about tripled to almost $15,000 in 1913, its net income did not grow as quickly. Annual expenditures for grants thus remained under $300 most years, and at times sank to $110. Throughout this period—when assistant professors at Columbia University in the nation's most costly city earned about $1,200 annually—the association paid Howard about $1,500 for his part-time job.[38]

Other problems persisted. Annual AAAS *Proceedings* volumes appeared months after each annual meeting, duplicated much that had previously been published in *Science*, and cost the association about $1,900 annually. No wonder reformers in 1896 tried to reduce their size and cost! In December 1907, the committee on policy again recommended that the *Proceedings* include "simply [the] list of members, constitution, and business of the meeting." But its motion barely survived a council attempt to table it. Four years later, Minot even called for the *Proceedings'* discontinuation, and Cattell offered to circulate "the membership list . . . as a supplement to *Science*." This proposal proved too extreme for their colleagues on the committee on policy, who agreed, however, to support simple triennial "lists of officers and members" and quadrennial volumes of *Summary Proceedings*. Whatever their form, these volumes rarely appeared expeditiously, and Howard continually reported delays to the council and committee on policy.[39]

These changes did reduce costs, however, and allowed the committee on policy to begin work for the association's growth. At the 1912–1913 Convocation Week meeting, for example, it authorized the appointment of a full-time associate secretary charged with "carry[ing] forward this extension work."[40] Although this work had little impact, other contemporaneous initiatives proved more successful. In that year, for example, the AAAS voted to meet in August 1915 in San Francisco. It had long considered a Pacific coast meeting, and the chance to meet concurrently with the city's 1915 Panama-Pacific International Exposition (with its reduced-rate rail fares) led finally to the vote. The thirty months between the decision and the event itself—unusually long for an early-twentieth-century meeting—allowed the Pacific coast planners, led by University of California astronomer William W. Campbell,

to organize themselves effectively. They began a well-considered membership drive in April 1913, and in May 1914, at its annual meeting in Seattle,
the Pacific Association of Scientific Societies—which had long supported
the region's scientists—took steps to reconstitute itself as the Pacific Division of the American Association for the Advancement of Science, the
AAAS's first regional division.[41]

Though Howard remained as passive as ever, MacDougal and Cattell
actively assisted in the planners' activities. Cattell advised on such practical
matters as the content and design of membership-recruitment material, and
had copies of *Popular Science Monthly* and *Science* sent to potential West
Coast members.[42] Most important, he publicized their plans and the general
topic of "Science on the Pacific Coast" in his journals. In March 1915, he
published and distributed widely an especially well-illustrated Pacific coast
issue of the *Monthly* that featured articles on the region's botany, entomology, geology, and paleontology, on "the last wild tribe of California," and on
Western astronomical observatories and biological laboratories.[43] But these
efforts attracted fewer eastern and midwestern scientists to San Francisco than
all had hoped. After all, despite faster trains and reduced-rate rail fares, transcontinental travel was still expensive and took several days. The efforts did
secure, however, the future of the association's Pacific Division, which began meeting annually in 1916 and remains today the AAAS's largest geographically based division. In 1920, again with MacDougal's leadership,
scientists in Arizona, New Mexico, and western Texas organized a southwest
division that immediately proved active as well.[44] These new divisions focused almost entirely on their regions' professional scientists and their interests, and thus served both to strengthen their communities and to solidify the
AAAS's definition of "advancement" as an inward-looking process.

Expanding and Competing in the 1910s

The Committee of One Hundred on Scientific Research as an Advocate for Science

The creation (in April 1913) of the Committee of One Hundred on Scientific Research embodied an attempt to implement more fully the AAAS's
umbrella function, to extend its concern for interdisciplinary links among scientific specialities, and to advance science by promoting research in all fields.
Through the late 1910s, it occupied much of the AAAS leaders' attention.
But though they created the body to address issues of central importance to
the AAAS's goals, the committee on policy apparently never recognized the

structural weaknesses of the earlier Committee of One Hundred on National Health. Its members merely hoped, instead, that this new body would help the association bypass Howard's stifling impact. In doing so they named the association's retiring president, sixty-seven-year-old Edward C. Pickering, as the chairman of its executive committee, and Cattell as its secretary. Pickering had directed the Harvard College Observatory since 1877 and, in serving for years as trustee of two small Boston-based endowments, had long used connections with Boston's social elite to build financial support for scientific research. The association's leaders hoped that the Committee of One Hundred would profit from Pickering's experience. But they never clearly defined its charge, just as in previous years ways of implementing (and even defining) the AAAS's overarching goals often remained unclear to many.[45]

At its December 1913 meeting in Atlanta, the council took the practical step of instructing Howard and Cattell "to prepare a directory of funds available for Research work." During the long train ride north after the meeting, Cattell, Pickering, and retiring AAAS president Edmund B. Wilson (a Columbia biologist) tried to give further shape to the Committee of One Hundred's ambiguous charge. But at Pickering's insistence, they concluded all too vaguely "that this committee was established in the hope that it might exert an influence favorable to the promotion of research work under the Government and in our universities by acting as an advisory body and in other ways endeavoring to cooperate toward the end indicated." Others agreed that though "the idea seems to me to be a good one, . . . [the] method of developing it to a practical finish is rather vague," and the AAAS's commitment to the committee's amorphous structure prevented its leaders from focusing its concerns. Cattell's form letter to AAAS members appointed to the committee called just as vaguely for "suggestions in regard to the work the committee might undertake to advance scientific research in America and to improve the conditions under which it is carried forward."[46]

Replies to this letter were equally imprecise. They urged the Committee of One Hundred to assess "possibilities in the way of coordination of existing activities," to "call attention to excellence in performance," and to demand higher salaries for college professors. The leaders did create several subcommittees—on research funds, on research in educational institutions, on industrial research, on the selection and preparation of research students, and on honors for noteworthy researchers—that promisingly focused on issues of significance for all sciences and, especially, on their interplay. But after deciding that "the principal work of the committee should be entrusted to subcommittees," the committee's leaders never provided much direction, and called simply for large-scale support for science without restrictions. The committee thus served to allow researchers to air personal grievances, and at its second meeting in December most of its subcommittees reported regret at having done so little. Cattell himself polled university administrators on the "weight . . . given to scientific research and productive scholarship . . . in making appointments and promotions" but early in 1915 admitted to Pickering that "university work and other matters" had led him to slight Committee of One Hundred activities.[47]

The committee's uncoordinated efforts to define and implement "advancement of science" thus sputtered along. After one meeting, Willis R. Whitney, General Electric's director of research, wrote that "twelve year old children can do more constructive work in an hour than a committee of one hundred scientists could do in a year."[48] Despite its leaders' hopes and after serious consideration, the Rockefeller Foundation rejected the committee's request for $50,000, to be awarded in small grants to AAAS members "likely to use the money to advantage in scientific research."[49] Some of its subcommittees did complete several potentially useful projects. For example, though the Pacific coast subcommittee called as vaguely as its counterparts for public recognition of science, it also urged Western universities to support sabbatical leaves with precisely defined benefits.[50] Most notably, the chairman of the subcommittee on research funds, MIT physicist Charles R. Cross, moved swiftly to prepare the "directory of funds available for research work" that the AAAS council had called for, which *Science* published in installments between 1916 and 1918. As Cross emphasized, his "list of endowments" and other sources of support was "doubtless incomplete." But this general roster—soon supplemented by lists of support available for medical research and at American colleges and universities—comprised twenty-seven separate funds managed by sixteen different agencies and foundations, and he "believed [it] to comprehend the more important." And as the list named the individuals to whom "application should be addressed . . . for information regarding the conditions upon which grants may be made from any particular fund," this directory proved to be the Committee of One Hundred's earliest, most substantial, and probably most useful product. Not coincidentally, it also was in many ways the result of the committee's least narrowly focused project.[51]

In the meantime, after a long legal battle, the association reached accord with the other heirs of Richard T. Colburn, a corporate financier and former vice president for the section on social and economic science. In April 1916, it finally inherited about $75,000 as the "Colburn Fund for research in the physical or psychic demonstrable sciences." This bequest quadrupled the Association's endowment, and Cross's directory noted that "regulations regarding [its] use [were] not yet formulated." Debates delayed their design, and finally, early in 1917, the AAAS committee on grants divided its $4,000 appropriation from this fund among its disciplinary subcommittees in accordance with the proportion of researchers in each field listed in *American Men of Science*. This decision limited the funds available to each subcommittee, so that none could award even relatively large grants, and several scientists dropped plans to submit applications. Some fields had access to only $300, and their subcommittees had to decide whether to award one grant of $300 or three of $100. As the grants committee's spring meeting neared, Cattell told Pickering that the subcommittees had not yet made "sufficient desirable recommendations to use the amount of money available." The final list of grants included several in most scientific fields, but totaled less than $4,000. And in 1918, in again appropriating $4,000 for the committee on grants, the AAAS committee on policy explicitly urged the committee to "use this amount with especial care."[52]

Competing with the National Research Council

Whatever its faults, the Committee of One Hundred represented the diverse interests of the "democratic" AAAS. But as the committee emerged in 1913, the more explicitly elitist National Academy of Sciences also took steps to enlarge its influence. In that year, George Ellery Hale (Director of the Carnegie Institution's Solar Observatory on Mount Wilson) began publishing in *Science* a series of articles on "National Academies and the Progress of Research." These concluded in 1914 with an especially influential article on "The Future of the National Academy of Sciences" that reasserted its role as scientific advisor to the federal government. At the Academy's April 1916 meeting, Hale cited recent events in Europe to call for it to reclaim this role immediately. The Academy then offered "to organize the scientific resources of the country in the interest of national preparedness." Once President Woodrow Wilson formally accepted this offer, Hale took charge of a committee that organized the National Research Council, "whose purpose shall be to bring into coöperation existing governmental, industrial, and other research organizations" in preparation for America's expected intervention in the European war.[53] The council soon became the "operating arm" of the Academy, through which it exerted a powerful influence on American science, its practitioners, and their organizations, including the AAAS. And to nobody's surprse, this influence reflected both the institutional tensions in the American scientific community between its "democratic" and its "elite" branches, as well as the personal alliances and working relationships developed by the leaders of these organizations.

Hale's and the Academy's plans thus had serious consequences for the AAAS's Committee of One Hundred. Many leading American scientists belonged to both groups, and both sought to promote the scientific community and secure greater support and a larger social role for scientists and their organizations. But many also had more faith in Hale's leadership (and ties with the federal government and its bureaucracy) than in Pickering's and Cattell's. That July, as both groups lobbied to create state engineering experiment stations (much like agricultural experiment stations), some committee members urged the AAAS to allow Hale's new council to lead the campaign.[54] The next month, *Nature* published Hale's detailed plans for the council even before *Science* received a copy. These plans emphasized preparedness and support for the allies and won support from most American scientific organizations and their members, including most AAAS leaders.[55] Most notably, in September 1916 the council began to form "joint committees on research in various sciences . . . in cooperation with the corresponding national scientific societies." It thus duplicated the Committee of One Hundred's disciplinary subcommittees, and later that month Hale invited the AAAS to "discuss measures by which the AAAS may cooperate in the research council work." By that time, Hale's plans for the NRC had developed so fully and had grown so in stature that the AAAS committee on policy waived precedent and named such a committee "without awaiting the meeting of [its own] Council." Late in November, the joint Committee on Cooperation called for the two organizations to merge their "committees in the various branches of science, with the understanding that the present members of the Association committees will become members of these committees of the National Research Council."[56]

Hale soon clarified this proposal by describing it as one "by which the Academy, the Association, and the National Scientific Society in the particular field involved, will each appoint one-third of the members" of the council's committee for each science. But most knew that it essentially called for the NRC to co-opt the Committee of One Hundred, its subcommittees, and all of their functions except the award of AAAS grant funds. Cattell strongly denounced this proposal. He had long opposed preparedness (from 1914 his *Popular Science Monthly* denounced both Allied and German militarism), and he resented Hale's attempts to co-opt a AAAS initiative. Matters came to a head at the December 1916 meeting of the Committee of One Hundred, at which Cattell's "spiteful" attack on Hale cost him support. As Harvard geographer William M. Davis telegraphed to Hale that evening, Cattell "unpleasantly attacked [the] aristocratic National Research Council [as] essentially Hale [and MIT chemist Arthur A.] Noyes and urged [the Committee to] refer [the] question [to the] democratic Association Policy Committee[. Clark University physicist Arthur G.] Webster replied amusingly[,] saying [that the] Policy committee is Howard [and] Cattell." At final count, only two of the fifty-one other committee members present voted with Cattell, and the Committee of One Hundred soon became moribund.[57] Twenty years after Cattell's revival of *Science* had also helped revive the AAAS, his personality and vision for the AAAS began to limit the association's ability to play a leading role in the scientific community.

Planning for the Future and Responding to the War
The National Research Council's success scared many of the AAAS's leaders; MacDougal even claimed that it threatened "the very welfare of the Association." It thus made pressing deliberations about the association's future that had begun months earlier. These aimed both to strengthen the association's ties with its affiliated societies and to provide stronger executive leadership, explicitly like Hale's of the NRC. The formal plans that emerged late in 1916 called for Howard's transfer to the honorary office of "Secretary of the Council" with nominal duties and, more important, for the redefinition of the position of general secretary, who had served largely as a recording officer. Under this plan, the new general secretary would be charged to be "active in the work of the organization of the association in its relation to the affiliated societies and the general scientific activity of the country." Indeed, some members of the committee on policy began to speak of a "federation of scientific organizations" as reorganization's ultimate goal.[58]

In December 1916, the council formally named a committee (comprising Cattell, Fairchild, and MacDougal) to embody these plans in a proposal for a revised constitution, and elected Cattell as general secretary. He was then fifty-six years old and planned to resign his professorship to assume a full-time, paid position as the association's chief executive officer. In looking ahead to re-creating the association as one of "societies rather than of individuals," Cattell envisioned a vastly richer AAAS that could well afford to pay a full-time executive the $10,000 that MacDougal (at least) suggested. MacDougal, however, also knew that not all AAAS leaders shared his admiration of Cattell, or these goals, or his resentment of the National Research Council. He identified this "conservative element" with the federal scientists who slighted the interests of their academic colleagues. And such Washington-based AAAS influentials as Howard and Weather Bureau physicist W. J. Humphreys did force the committee to delay action.[59]

The nation's entry into World War I in April 1917 distracted most American scientists and further changed the AAAS's institutional environment. The National Research Council coordinated science's role in the war effort, and for the most part its "joint committees" for the various sciences operated without reference to whether their members were appointed by the National Academy, a specific national scientific society, or the AAAS. And as the NRC began operating as the "Department of Research" of the Council of National Defense, it soon downplayed its discipline-based structure, organized its activities more functionally, and established committees charged with overseeing developments relating to agriculture, aeronautics, food, noxious gases, navigation and nautical instruments, and the like. This functional organization left some discipline-based committees with little or nothing to do, and early in 1918, for example, the AAAS section on botany called for "the immediate mobilization of the botanists of the United States and for their more effective co-operation in the prosecution of war work." The AAAS committee on policy, however, knew who had power and referred the request to the "Agricultural Committees of the National Research Council, with a request for active co-operation and assistance."[60] The functional arrangements that evolved thus served both the war effort and the Research Council well. But the war had no real impact on either the AAAS or even the National Academy, both of which retained their discipline-based sections and governance schemes. Indeed, throughout the war the AAAS itself became almost moribund, and its leaders seriously considered canceling its December 1917 meeting. When they eventually held the meeting in Pittsburgh, it drew almost as few attendees as had the 1915 West Coast meeting.[61]

Earlier that fall, however, the association had to deal with an especially serious problem of governance and administration, after Columbia University dismissed Cattell from its faculty for his opposition to aspects of U.S. conscription policy. Though some of his closest friends and the liberal press claimed that the firing violated Cattell's rights to free speech and academic freedom, most American scientists defended Columbia's action. At the same time, although one or two Washington-based scientists hoped to use the incident to pry *Science* from his hands, many American scientists urged Cattell to remain as editor and urged the association to retain *Science* as its official journal. As long-time AAAS treasurer Robert S. Woodward wrote, "I am disposed to condemn [Cattell's] politics roundly. I think him not only quite wrong but very unpatriotic. But this subject may be kept quite apart from his field as editor of *Science*." Quite simply, most American scientists outside of Washington (and even some in the capital) believed strongly that he had to stay as *Science*'s editor, and many urged the committee on policy to reaffirm his editorship.[62]

Cattell did resign as AAAS general secretary, as secretary of the Committee of One Hundred, and from the constitutional revision committee. But despite his dismissal's major impact on most aspects of his life, it had little impact on his editorship of *Science* and his other journals. In November 1917, at Howard's urging, the AAAS committee on policy reasserted that *Science*'s editorial board included its members and the chairmen of the various AAAS sections. But though this committee formally charged this board to "control" *Science*, it never did. In the years that followed, the committee's successors tried at times to claim this editorial board's authority, but Cattell regularly ignored it, even as he never openly challenged its recommendations. And in 1919, as all realized that the AAAS had finally to confront postwar inflation, Cattell convinced the committee on policy to amend the constitution to raise annual dues to five dollars in order to pay him three dollars for each annual subscription to *Science* he supplied to AAAS members. This was the first raise that Cattell had received in the twenty years that *Science* had been the AAAS's official journal.[63]

Thriving in the 1920s

The war ended abruptly in November 1918, and in the Roaring Twenties, large numbers of Americans felt richer than ever. The economy that boomed for many (at least until October 1929) reinforced the growth of the middle classes, and with that growth came new vitality in universities and colleges. On their campuses, science-based technologies—especially chemical engineering, electrical engineering, and metallurgy—increasingly attracted students, and the industries based on these technologies thrived. Fortunes previously created by earlier science-based industries (like petroleum and steel) continued to feed the philanthropic foundations founded early in the century, including the Rockefeller Foundation and its associated funds and the several Carnegie philanthropies.[64] As public interest in the sciences grew—stimulated in part by the introduction and rapid growth of broadcast radio, itself clearly dependent upon science—these institutions increasingly offered support for scientific work. The public may not have fully grasped the explication of the new genetics by T. H. Morgan and his students, or the complexities of relativity or quantum or cosmic-ray physics. But early in the decade, Albert Einstein became a nationally recognized celebrity familiar to all American newspaper readers and radio listeners.

Completing Reorganization for Stability and Growth

Like many organizations, the AAAS worked to take advantage of this sense of wealth and these other developments, even as it adapted to a new scientific environment in which the National Academy and its National Research Council had become dominant. In the new environment, the AAAS could no longer assume that it defined science, led the coordination of disciplines, or even served as science's primary advocate in the federal government. Except for its ties with *Science*, which continued to serve as a forum in which the scientific community could explore both new findings and its own advancement, immediately after the war, the association found itself shut out of the ecological niches it had previously occupied. Attempts in the previous decade to win support from one or more of the Carnegie or Rockefeller philanthropies had failed, and even its annual national meetings faced challenges to their survival. In 1918, for example, AAAS leaders realized that Washington had "become the center of scientific activity of the country" and moved that year's meeting from Boston to Baltimore. Despite this change, the meeting drew even fewer attendees than had the 1917 Pittsburgh meeting. AAAS influentials like MacDougal and Cattell thus knew they had to take steps in the war's aftermath to resuscitate the association and its programs, to create new niches. MacDougal had long complained that Howard had focused his attention on Washington issues exclusively, and he began criticizing the growing impact of the "Washington element" both on the association and on American science at large. To him the Washington community seemed to "tak[e] the attitude that it is up to them . . . to guide and direct the destinies of the country in all of its details." Reflecting the AAAS's traditional democratic commitments, he argued that the association had to counter these attitudes.[65]

The primary goal of MacDougal and his allies remained replacing Howard as the association's chief executive. But they knew that the AAAS could not pay a full-time salary, and most of the scientists with executive experience who were considered took instead well-paid presidencies of universities or foundations. In 1920, the association finally named plant physiologist Burton E. Livingston as its permanent secretary. Only forty-five years old, Livingston had built a distinguished career that included the editorship of *Botanical Abstracts* and extensive research at, successively, the U.S. Department of Agriculture's Bureau of Soils, the Carnegie Institution's Desert Botanical Laboratory (under MacDougal), and Johns Hopkins, where he had been a professor since 1909.[66] He agreed to devote one-third of his time to AAAS affairs and to commute two or three days a week from Baltimore to the association's offices at the Smithsonian Institution for an annual salary of $2,500. For the next decade Livingston provided the AAAS with the energy and stability it had long sought to promote its growth and better advocate for its members and their work.

Once in office, Livingston found that the association's rooms at the Smithsonian were littered with "bottles and corkscrews," that much AAAS office equipment had been stolen, and that its one full-time employee, Assistant Secretary F. S. Hazard, often drank on the job. (No wonder membership records, which Hazard oversaw, proved rarely accurate.) Livingston fired Hazard, and in his place hired as executive assistant Sam Woodley, who soon brought some order to the chaos. Woodley remained with the AAAS for the next twenty-five years, and often proved the most stable feature of the AAAS's administration. In December 1920, Livingston published in *Science* a report on the "Present Status of the Affairs of the American Association for the Advancement of Science" that for the first time provided its members with a notion of his office's practical responsibilities. It also outlined a renewed vision of the AAAS as a democratic institution that could best advance science by representing all scientists and their specialized organizations. He described plans to increase the association's influence and expand its role as an advocate by further extending its organizational purview.[67]

The constitutional revision that the council adopted earlier that year embodied these goals by calling for closer relations between the association at large, its affiliated societies, its regional divisions, and its as-yet unorganized local branches. It thus added to the constitution's "Objects" statement, "to co-operate with other scientific societies and institutions." It also retained representation of all affiliates on the AAAS council in proportion to the number of AAAS fellows each had among its members. One of the association's first initiatives under its new constitution involved recruiting formal ties with those societies that had not previously been affiliates. By 1925, the AAAS listed fifty-three affiliated societies, sixteen affiliated state academies of science (which also had council representation), and more than thirty associated societies. Early in the decade the AAAS's formal letterhead read "American Association for the Advancement of Science and Affiliated National Scientific Societies." Though many thought the claim embodied in this letterhead overstated the AAAS's actual influence, it did represent an important goal of community building that—through the 1920s—the association could reasonably justify.[68]

The new constitution also realigned the disciplinary identifications of some sections and added others for "Agriculture," "Historical and Philological Sciences," and "Manufactures and Commerce." And from 1920, the restructured council also included representatives from all AAAS sections. Since individual AAAS members elected these delegates, Cattell and MacDougal claimed that their scheme implemented the association's commitment to democracy. In reality, the new constitution further centralized AAAS governance and placed much power in the hands of a new executive committee, which replaced the committee on policy. This committee comprised the association's officers and eight other members serving four-year terms. All were elected by the council, and these procedures embodied the representative (rather than direct) democracy that *Science* had praised in the 1890s. Under the new constitution, AAAS members chose their representatives and then deferred to their judgment. And as the executive committee evolved, its members arranged things so as to control AAAS affairs completely.

Without today's concern for diversity, the council continued to look to the middle-aged white men who had long controlled AAAS affairs. The first members elected to the executive committee included all the usual suspects, such as Cattell, Fairchild, Humphreys, MacDougal, and Henry B. Ward, all of whom had served on the committee on policy. All were fifty-five years old or older, and they effectively dominated AAAS affairs. As one member finished his term, his colleagues immediately arranged his reelection. Late in 1925, they further consolidated their control when they had the council adopt a bylaw granting the executive committee "full power to act for the Council when the Council is not in session." The executive committee always did include a few new faces, well-known scientists elected annually by the council in uncontested elections to the AAAS's presidency, including paleontologist Charles D. Walcott (elected in 1923) and horticulturist Liberty Hyde Bailey (1926). But through the 1920s, the executive committee always chose the one presidential candidate. At least a few presidents accepted election solely for the honor, many of them missed most executive committee meetings during their terms, and most remained unfamiliar with AAAS policies and procedures during their single year in office. In at least one case a president (eminent University of Chicago mathematician Eliakim H. Moore) had not even been a AAAS member until just before he was elected in 1920.[69]

In the same way, the association's ever more gerontocratic leadership typically restrained the younger members who at times joined Livingston on the executive committee. In 1922, for example, Livingston himself wished for "new blood" on the committee but knew that, since "the 2 to retire" that year were Cattell and Ward, "they'll both be returned." He and others criticized the committee because it never "let the members know just what we are doing." Livingston also claimed that "I don't think it is right & best for the AAAS to elect its president by what looks like secret diplomacy," and suggested several new procedures. Most seriously, he argued that "I don't think we should leave the choice mainly to Cattell." But he knew where the power lay and concluded that "I'm not going to do anything about it . . . unless I am sure of support."[70] Late in the 1920s, the executive committee did begin asking AAAS members to submit names of possible presidential candidates, and in the 1930s, successive councils received slates of nominees from which its members actually elected a president.[71] But more fully democratic election procedures that had the AAAS membership elect presidents and other officers emerged only in the 1970s.[72]

As Livingston's complaints implied, by the mid-1920s, Cattell had begun to treat the AAAS as his personal fiefdom, and he sought to direct all AAAS affairs. Since Columbia University no longer demanded his attention, he focused more and more of it on the AAAS, even though (as perhaps only MacDougal realized) he spent less and less time actually editing his journals. His forceful personality and self-assurance also stifled any tendency to question his views. In 1925, despite the executive committee's 1920 vote that "the term of office of the chairman [shall] be no more than four years," its members confirmed his dominance and elected the then sixty-five-year-old Cattell as "Permanent Chairman of the Executive Committee," an office he held through 1941, when he turned 81.[73]

Promoting Science Through Reinvigorated Annual Meetings

To many, of course, the association's primary function remained its annual meetings, and after the problems of 1917 and 1918, the first postwar meetings proved especially important. These took place in a new political climate of concern about communism (exemplified by the Red Scare, the Sacco-Vanzetti case, and the creation of the FBI), and AAAS influentials sought to insure that the association and *Science* would not become politicized. Permanent Secretary Howard complained about a 1919 *Science* article that attacked the gold standard, the private ownership of land and other resources, and federal taxation policy in the name of "The Scientific Spirit." Later that year, speakers addressing several sections at the association's St. Louis meeting claimed that "science" supported their positions that the United States should (or should not) join the League of Nations. Hoping to prevent similar situations, the committee on policy urged "that sectional officers avoid placing on their programs papers relating to political questions on which public opinion is divided." Many St. Louis newspapers supported this resolution and the association's leaders believed it would prevent further difficulties.[74]

In April 1927, the AAAS went even further. In the aftermath of the trial of John T. Scopes for teaching evolution—and despite the council's earlier resolutions supporting Scopes and denouncing the laws that led to his conviction—in planning for the December 1927 annual meeting in Nashville, the executive committee argued "that special programs on evolution and unusual emphasis on this subject should be avoided." It did note, however, that "such programs and such special emphasis may well be arranged" for the next year's meeting, scheduled for New York. But at the same time, it even asked the affiliated "societies meeting with the Association . . . to exclude from their programs . . . papers that might be regarded as too sensational."[75] In less than a decade, however, as the AAAS and its members had to deal with the Great Depression, the association reversed this position, thus beginning the move toward concern with social issues that flourished in the years after World War II.

In promoting postwar meetings, the AAAS tried to revive the spirit of Convocation Week, even as its leaders knew that many affiliates would not agree to join all such gatherings. By 1922, they had evolved a scheme that called for large Convocation Week meetings every leap year in Washington, New York, or Chicago; for smaller meetings in the intervening even-numbered years in other large cities to which "national scientific societies" were invited; and for meetings in odd-numbered years in other cities, to be arranged "for the convenience of the scientific societies that may find it desirable to meet with the association." These recommendations also called for "more informal" summer meetings at "universities and places not suited to a large winter meeting," and implied that the regional divisions would organize them. Before 1926 the association held summer meetings in Salt Lake City, Los Angeles, Boulder, and Portland, Oregon. These proved scientifically successful, and the Salt Lake City meeting featured, for example, a major symposium concerning the Colorado River. Though the association never implemented all aspects of this plan, it set the outline of a policy that served its members well through the 1930s.[76]

In many ways, this scheme derived from the great success of the 1920 meeting in Chicago. Almost twenty-five hundred attendees registered with the AAAS, many hundred others registered with 41 affiliated societies, and the official program listed over a thousand papers. Even the American Physical Society met with the AAAS that year, and University of Chicago physicist Albert A. Michelson, then still America's only Nobel Laureate in the subject, greatly impressed his audience with his report of his measurement of the diameter of the star *Betelgeuse*. Arthur Holly Compton, then professor at Washington University, seven years later won a Nobel Prize himself, but he remembered Michelson's address and the meeting itself as an early high point in his career.[77]

So too did Ernest E. Just, the African-American embryologist who had become a professor at Howard University long before he earned his University of Chicago Ph.D. in 1916. His audience at a 1920 symposium on fertilization received his remarks especially well. Perhaps more significantly, attending the meeting gave Just what his biographer called "a sense of professionalism and camaraderie he could scarcely have found elsewhere." During and after the meeting, Just grew to see himself a part of the larger scientific community and felt himself "inspired with a [new] sense of purpose."[78] The Chicago meeting thus demonstrated how the AAAS could serve to establish and maintain a sense of community among American scientists spread across the nation's breadth and—as social, racial, and ethnic sensibilities changed during the course of the century—across its human diversity as well.

Other AAAS meetings in the 1920s had equally strong impacts on members even more senior than Just, or even Compton. In 1925, for example, Columbia University electrical engineer Michael I. Pupin served as president, and his remarks to council emphasized the meeting's "enthusiasm and inspiration," which he urged those present to "carry back to those who were unable to attend." More generally, he called on AAAS members to "preach the gospel of science to the people at all times."[79] Compton's, Just's, and Pupin's cases thus provide major examples of the continuing impact and significance of the AAAS's original purpose of establishing a community of science.

Despite such successes, not all AAAS sections held such influential sessions through the 1920s. The section on manufactures and commerce never organized itself, and in some years other sections did not even sponsor vice-presidential addresses. For several years, officers of the section on historical and philological sciences argued for its redefinition as a history of science section, an argument that the council would not even consider.[80] The executive committee also always feared that the biological societies might stop meeting with the AAAS. If they did, Livingston projected, the biological sections would have trouble organizing sessions, and the association at large would lose about "half of our attendance and program."

Though Livingston never expected this outcome, the possibility sparked musings about how the AAAS might further promote the advancement of science, and in 1922, he began speculating about future AAAS meetings. As he asked MacDougal, "Might the AAAS not take the broader phases of science and the matter of educating the public as its main chore? As fast as our sections are left without any societies meeting with them, we might introduce section programs of more personal & popular nature. . . . Our meetings might become of much more general interest, not so technical."[81]

Livingston's thoughts did not include a return to the public lectures that AAAS meetings had highlighted through the 1890s. They ran, instead, to broadly focused nontechnical presentations by scientists about their own work addressed to scientists in other fields, much like the articles published in the *Scientific Monthly*. At that year's annual meeting in Boston, Cornell president Livingston Farrand gave such a talk, as the first of an annual series of lectures sponsored until the 1970s by Sigma Xi, the honor society of science. Throughout the 1920s, the executive committee regularly sought to schedule other analogous lectures and urged section committees to do the same. Livingston especially had hopes that other organizations would follow Sigma Xi's lead.[82] But despite some sporadic successes, most programs at AAAS meetings through the 1920s and 1930s remained as technically focused as ever.

More prosaically, increased meeting costs led to proposals for a meeting registration fee of $2.50. This suggestion proved too radical for the executive committee, which voted instead to charge a fifty-cent "ticket endorsement and validation fee" for those seeking reduced-rate rail fares.[83] More significantly, in the 1920s, the AAAS converted the previously ad hoc exhibits held in conjunction with its meetings into formal Annual Exhibitions of Science. These combined commercial exhibits (sponsored by manufacturers and dealers of scientific apparatus) and "research exhibits," presented by "individual scientists and scientific institutions and laboratories." The exhibitions gradually grew in size and significance, and proved especially influential as in many ways the research exhibits served as the new organic form from which evolved the poster sessions that dominate many late-twentieth-century scientific meetings. In the 1920s, however, tensions between the exhibitions' commercial and educational functions soon emerged. In seeking to make money from them, the executive committee in 1925 appointed as exhibition manager a salesman with no scientific credentials and promised him half of the

Exhibition's net profits. With apparently little appreciation for the conduct of science, by the end of the decade he had so alienated nonprofit exhibitors that, in 1930, the executive committee assigned the research exhibits to the AAAS committee on research and ordered that "a rail or fence" be used to separate them from the commercial exhibits. For many years thereafter a committee on research exhibits chaired by F. C. Brown (a physicist at the National Bureau of Standards) oversaw this aspect of AAAS annual meetings.[84]

Science *and the AAAS in the 1920s*

As reorganization efforts bore fruit, MacDougal and Livingston turned to *Science*, which, they believed, did not serve American science (or represent the AAAS) as well as it should. Their concerns remained largely undefined but emerged in part because, in the 1920s, *Science* closely resembled (in style, format, and even in the kinds of material featured) the journal that Cattell had first edited in the 1890s. They had to admit that *Science* still published major scientific news and articles, such as notes by Einstein and Clinton H. Davisson's early reports of his electron diffraction experiments. After all, ambitious scientists had long since learned to take advantage of *Science*'s ability to publish significant research results rapidly, especially when first presented at AAAS meetings. But MacDougal and Livingston also knew that more and more of their colleagues were publishing other major papers—especially those that extended and explicated the implications of early laboratory studies—in their own specialties' most prestigious journals. In this way, for example, Davisson published the full report of his electron diffraction studies in the *Physical Review*.[85] Livingston and MacDougal also believed that *Science* wasted valuable space by publishing in full the often purely ceremonial addresses given at annual meetings by each section's vice president. They also had the vague sense that *Science*'s "Scientific Notes and News" downplayed the increasingly noteworthy news of the "Washington element."[86]

More definitely, they knew that *Science* did not support Livingston's AAAS membership recruitment efforts and never carried advertising for the association, even in issues sent only to nonmembers. For Cattell, the financial value of adding new members had changed from the century's first years. He began claiming that the many libraries that (and the few individuals who) paid *Science*'s full subscription price (as opposed to those who received it through AAAS membership) made the journal financially possible. He thus feared a loss of their subscriptions. As a result, in Livingston's words, "we try to get members in every way *except to advertize* [sic] *in our official journal.*" As he continued, "It seems to me the whole story about being a

member should appear in every issue of *Science*" and that "we must lead
Cattell to this pond and get him to drink." But he had to admit that "I haven't
any clear idea as to just how it is to be done."[87]

This problem never stopped MacDougal and Livingston from discussing how they would alter *Science* if they had the chance. In December 1921,
for example, MacDougal formally called for the executive committee to in-
struct Cattell to delete from all formal addresses published in *Science* all in-
troductory remarks "designed [solely] to bring the speaker and the audience
together." The committee debated this proposal for hours, and ten months later
its members finally voted that these papers should appear in *Science* "without abridgement."[88] The following year, Livingston suggested that "*Science*
might be more useful to the scientists if we publish more technical papers,
brief ones, generally preliminary announcements." MacDougal agreed: "Of
course the things you mentioned and others would improve *Science*, but how
can we make any changes in *Science*?" As he concluded, "I do not expect
much, if any, change in *Science* under our present arrangement."[89] In December 1922, however, after MacDougal again called attention to *Science*'s
weak book review section, the executive committee agreed to consider "the
need of more and better American reviews of scientific publications." In the
months that followed, MacDougal and Livingston speculated on the future
of what they began calling *Science Reviews* and on its relations with *Science*.
But after some intense discussion, the initiative died by the middle of the
year. Equally short-lived were Livingston's plans for a quarterly bulletin publishing AAAS news and announcements of forthcoming scientific meetings
that, he hoped, would bring favorable attention to the association. He still complained that *Science* never carried advertisements for AAAS membership, and
in 1924, during Cattell's own term as AAAS president, told MacDougal that,
"If you want to explode [sic] something [with Cattell] you might suggest
standing or intermittent advertizing. . . . But that's on the knees of the gods,
meaning largely in the hands of the president."[90]

In April 1925, after some discussion with other longtime AAAS
influentials and just before his sixty-fifth birthday, Cattell submitted to the
executive committee a proposal "to assure the ultimate acquirement of [*Science*] by the association, but to allow me to continue to conduct it so long as
I do so to the satisfaction of scientific men." It called for the AAAS to assume ownership of *Science* on his death or "relinquishment," subject to three
conditions: that *Science* remain the association's official journal; that the
AAAS pay his wife, Josephine Owen Cattell (who had long worked with him
in editing *Science*), a lifetime annuity of "one half the average annual net

profits of the journal" during the five years before the transfer; and that the AAAS continue to publish *Science* through his Science Press. All AAAS officials responded most positively to this offer, and Livingston immediately arranged for Roscoe Pound, dean of the Harvard Law School and (as a published botanist) an active member of the AAAS, to draw up a formal contract embodying these terms. By the end of July, Cattell and AAAS president Michael Pupin had signed the formal agreement committing both parties to its provisions, though neither the AAAS executive committee nor its council had had a chance to vote on it. In December, both AAAS bodies voted unanimously to accept what many saw as Cattell's extremely generous offer. Only Livingston worried about the agreement's potential impact on AAAS finances and on its other programs, and asked Cattell for estimates of the "average annual net profits of the journal," which he never received. Livingston and his colleagues also never had a firm notion of *Science's* editorial and production costs (which the association committed itself to pay once it assumed ownership), and Pound expressed concern about Cattell's third condition. But in the apparent financial security of the 1920s, even he concluded that "you really cannot look a gift horse in the mouth."[91]

The agreement had a major impact on all aspects of the AAAS. On one level, it led many to believe the association was in Cattell's debt and thus reinforced their tendency to defer to him.[92] Indeed, his election as the executive committee's permanent chairman occurred at the same meeting at which he first presented his agreement. Less personally, it reinforced the AAAS's caution in considering new programs, as its leaders knew that it might soon have to begin liquidating a major (and undefined) debt. By the end of the decade, even as AAAS officers feared the coming debt, they began looking forward to the time when they would be free of Cattell and what some already saw as his stifling impact on the association. As he turned seventy in 1930, they all believed they would not have long to wait.

Advocating Science Through Public Outreach and Other Programs

As the 1920s continued, AAAS leaders could cite as successes the association's revised governance scheme, the appointment of an active and energetic permanent secretary, regular growth in membership deriving from his continual efforts, the new southwest division, and reinvigorated, influential, and well-attended annual meetings. But Livingston and others continued to look beyond these programs, as they realized that Howard's earlier lack of leadership had "put the AAAS to sleep and we know how hard it is to awaken it after a Rip-Van-Winkle period." They thus promoted other initiatives that often broke new ground. For example, beginning in 1916, W. J. Humphreys had often urged the AAAS to sponsor a popular journal of science edited for the general reader, a proposal that lapsed once the United States entered World War I. He revived it in 1919, only to see it co-opted by plans for what emerged the following year (with support from newspaper magnate E. W. Scripps) as Science Service, a press service for science that proved especially influential through the 1920s. For many years, the AAAS and other organizations appointed members of the Service's board, and in the early 1920s, it provided a press bureau for several AAAS annual meetings. In 1923, Alva Johnston of the *New York Times* even won the Pulitzer Prize for his coverage of the previous year's AAAS meeting.[93]

In the decade's last half, however, Science Service devoted less and less attention to AAAS annual meetings, and late in 1925, members formally noted that "the publicity service . . . had not been as satisfactory as might be desired." Within six months, Austin H. Clark, a curator at the Smithsonian's U.S. National Museum, became AAAS director of publicity. Through the years that followed, Clark's efforts (and those of the AAAS Press Service that he formally established a few years later) brought much attention to the association and its annual meetings, and through the late 1920s, newspapers often reported on AAAS meetings in detail. By the decade's end, reporters from both the Associated Press and United Press frequented AAAS meetings.[94] The *Literary Digest* and other popular magazines ran "Glimpses of the March of Science" derived from the 1928 AAAS meeting and *Catholic World* emphasized that Robert A. Millikan's 1929 AAAS presidential address "rebuked . . . overhasty conclusions of a sensational nature" about the destructive effects of science and, in particular, denounced the claim that "'enormous stores of sub-atomic energy' could be used by irresponsible persons to wreak fearful destruction." Clark's publicity program also included radio talks on science on the NBC network, and through the 1930s *Time, Newsweek* and other popular magazines often reported AAAS news.[95]

Other AAAS programs began to address significant educational issues. For example, early in the 1920s, Livingston projected nationwide lecture tours by distinguished scientists designed both to bring the excitement of science before general audiences and to publicize the association and its programs. These tours, he claimed, would be "something the AAAS can do besides holding meetings and giving *Science* to its members." But he had to admit that his office could not manage the program. Later in the decade, Humphreys argued that such lectures might target high schools, and he stressed the "desirability of giving students in these schools a clear idea of the *method* of science." But he soon concluded that "it seems best not to attempt to develop this suggested feature of the work of the Association at this time."[96] In the years that followed, however, the AAAS established a committee on the place of science in education, long chaired by Otis W. Caldwell of Teachers College, Columbia University. Through the late 1920s, this committee prepared lists of recommended books for school libraries, proposed an essay contest of science for high school students, and planned "cooperative work with secondary schools, by which it is hoped to discover and encourage students who possess capacity for scientific work." But despite these continuing discussions, a lack of funding prevented action, and the AAAS's primary initiatives into direct science education took place in later decades.[97]

The AAAS also continued throughout the decade to try to promote science and influence public policy by adopting all sorts of resolutions at its annual meetings. In the early 1920s, for example, the AAAS campaigned against customs duties on scientific materials imported by universities, but this action alienated many of the American manufacturers who supported the National Research Council and limited their interest in the association's Annual Expositions of Science. Like other scientific groups, the AAAS drafted resolutions condemning antivivisection laws, advocating the adoption of the metric system and calendar reform, calling for income tax relief for scientists, and supporting conservation policies.[98] Perhaps most notably, in 1925 it joined with many other groups to defend Tennessee biology teacher John T. Scopes. Both Cattell and that year's AAAS president, Michael Pupin, met with Scopes to help plan his defense; the association's resolutions proved unusually unambiguous; and as articles in *Science* denounced the antievolutionary sentiments that led to the trial, Science Service press releases did much to spread evolutionary views. The AAAS itself had no formal presence at the trial, but many AAAS members joined with others to establish a fund to enable Scopes to continue his graduate study at the University of Chicago, and only a call for contributions addressed in 1926 to AAAS members enabled the American Civil Liberties Union to cover the cost of Scopes's defense.[99]

Even more productively, early in the decade, the association found the resources to develop a new mechanism designed to advance science by stimulating excellent research. In 1923, it established a "Thousand Dollar Prize," funded anonymously by Newcomb Cleveland, whose name it took after his death in 1951. AAAS leaders hoped the prize would attract younger researchers, especially in the physical sciences, and they thus defined it to recognize significant "research reported for the first time before some scientific session" at an AAAS annual meeting. Of the nine prizes awarded through 1930, eight went to papers in mathematics and the physical sciences. (The ninth went to H. J. Muller for a paper on "The Influence of X-Rays on Genes and Chromosomes," reporting work for which he later won the Nobel Prize in Physiology or Medicine.) Prizewinners through this period included (in 1923) thirty-five-year-old Edwin P. Hubble (G. E. Hale's colleague at the Mount Wilson Observatory) for his influential studies of spiral nebulae and (in 1930) twenty-nine-year old Merle Tuve and his collaborators at the Carnegie Institution of Washington's Department of Terrestrial Magnetism for producing high-energy beta and gamma rays. On the other hand, despite Cleveland's explicit call to reward younger scholars, in 1925, the association recognized fifty-nine-year-old Dayton C. Miller, Michelson's successor at the Case School of Applied Science, for his extension of the Michelson-Morley experiment that (he claimed) finally demonstrated ether drift and thus disproved the special theory of relativity. In 1927, as Cleveland renewed his support for these prizes, the committee described them as "one of the outstanding features of the annual meetings." The amount of each Newcomb Cleveland Prize rose to $2,000 in 1975 and $5,000 in 1976, the year before the association redefined it to honor the authors of papers published in *Science*.[100]

On a larger scale, in the 1920s, the association supported efforts to organize a federation of American biological societies. Many of these had long met jointly with the AAAS, and most American biologists saw no reason to look beyond these meetings to provide their sciences with some sense of intellectual and professional unity. Others disagreed, and in 1922, members of many of these organizations, including the AAAS's biological sections, began to meet to explore the matter. From the start, several AAAS leaders feared that what some began calling the Union of American Biological Societies might undercut the association, but in the 1920s, these concerns proved groundless, and the Union's organizing committee explicitly voted that "the aim of the federation would be to make use of the organization of the Association" and emphasized publications as its "main activity." This initiative led in 1926 to *Biological Abstracts*, but further unifying efforts proved unsuccessful, and biologists stopped speaking of a "federation" or "union." Through the late 1920s and 1930s, most biological societies looked as commonly as ever to joint AAAS meetings. The federated American Institute of Biological Sciences did not emerge until 1948.[101]

Despite these successes, some AAAS leaders remained jealous and distrustful of the National Academy and the National Research Council. As early as 1917, the council had won support from many industrial firms and foundations and, most significantly, from the Carnegie Corporation and its associated institutions. By the early 1920s, Carnegie support for the NRC totaled more than $5 million, and throughout the decade at least some AAAS officials envied and resented the council and its successes, and distrusted the motives of its leaders. For example, in response to an early offer of office space for the AAAS and Science Service in the National Academy's headquarters, Cattell wrote, "I expect that the National Research Council would like to eat us up, or at all events make us a subordinate branch of their organization." At the same time, MacDougal admitted, "I find it very difficult to prevent myself from getting into what is a constructively hostile attitude toward the National Research Council. The entire movement in its organization is toward superposing a system on us, in the determining of the nature of which system we had but little part."[102]

This envy and resentment emerged in a variety of ways. Livingston for one recognized that the NAS and AAAS shared the same goals—such as "bringing together scientists" and "encouraging an *esprit de corps* among scientific workers in general"—and worked hard to overcome this hostility.[103] But at the other extreme, Cattell drafted and widely circulated a humorous paper on "The Organization of Scientific Men" that contrasted what he described as the academy's and research council's "elitism" with the AAAS's commitment to democracy. It also snidely described the Carnegie Institution of Washington (which built the Mount Wilson Observatory) as George Ellery Hale's "greatest discovery" and charged that Carnegie Corporation president James R. Angell had sabotaged a AAAS grant proposal. Cattell's colleagues all appreciated his humor but urged him not to publish what they thought were libelous statements. Hale pointed out "certain inaccuracies" and emphasized the goals shared by the council and the association and their members: "I am sure we all agree in our desire to promote the interests of science and research." He thus asked Cattell, who also belonged to the National Academy, "Do you really think that the publication of this MS will help in this direction?" As published, the paper omitted Cattell's harshest claims, but it still described the academy's headquarters (built with Carnegie Corporation funds) as a "marble mausoleum," and its comparisons remained sharp and, to some, nasty.[104] In the decade that followed, other AAAS leaders found that Cattell's attitudes at times stood in the way of their best efforts to join Hale and others to "promote the interests of science and research."

Decline and Rebound through the 1930s and 1940s

Overcoming Organizational Instability

Like most American institutions, the AAAS suffered through the Great Depression of the 1930s, and its financial difficulties sharply limited its actions. In 1934, for example, when Harvard astronomer Harlow Shapley sought to extend his (and the association's) interest in promoting public understanding of science through a AAAS committee on adult education, Permanent Secretary Henry B. Ward told him that the association could provide only $100. Throughout the decade and the years that followed, many AAAS members devoted much time and attention to discussing a wide range of potentially significant programs. After World War II, many of these emerged in altered forms in ways that enabled the association to have a series of major impacts on the American scientific community. But before the war, at least, the AAAS could rarely act on its members' best ideas.

Another reason for the AAAS's inertia derived from its continuing inability to articulate its goals clearly, as it had failed to through the previous generation. This problem grew even worse in the 1930s, as Cattell solidified his control over AAAS affairs. As his journals' advertising income shrank during the Depression, he began to treat the AAAS more than ever as solely a subscription agency for *Science*. Those who had to work with him shared reports of the difficulties they encountered and expressed thanks that (so they told each other) his overbearance would not last long. But they also feared the potential impact of the AAAS's agreement to assume control of *Science* and to pay Josephine Owen Cattell all they had agreed to, and this fear further stifled the AAAS's freedom to act.[105]

Early in 1930, after Cattell "went out of his way to be troublesome" to Livingston, the permanent secretary announced plans to resign. He stayed on, however, until the end of 1930, when the executive committee—after deciding to name a younger permanent secretary "and let him grow" into the job—appointed economist Charles F. Roos, then secretary of the section on social sciences. But Roos was unfamiliar with the AAAS, and the central office staff, led by Sam Woodley, undercut his authority. Through 1931 and 1932, as the AAAS continued to drift, office routines collapsed, and membership declined. Rather than attribute these losses to the Depression, Cattell blamed Roos and had him fired.[106] By the middle of 1932, the executive committee arranged for Henry B. Ward, then sixty-seven, to become AAAS permanent secretary the following winter. Ward had just retired from the University of Illinois, where he had earned an international scientific reputation in parasitology and had built a significant administrative career as secretary of Sigma Xi from 1904 to 1921. He also knew the AAAS inside and out, having served on the council and the executive committee for many years. But he too encountered problems, as both Woodley and Cattell undercut his authority as they had Roos's. Executive committee member Edwin B. Wilson recognized the difficulty of working with Cattell—"I don't know anyone in the country who can work with him"—and even wrote that "I don't even know whether [AAAS affairs] can be worked out satisfactorily as long as Cattell has the prominent place he has."[107]

Most seriously, Wilson expressed particular concern about "the long-range policy of the AAAS and [even] whether it has any" and complained that "I have never got a picture of what we have to look ahead to." He once wrote—perhaps too harshly—that the AAAS had "followed an opportunistic career without program or policy but merely with a hope . . . that someday if we were patient enough and tactful enough . . . we might succeed in bringing [it] into a prominent position."[108] He and Ward tried in their own way to look ahead by recruiting into the AAAS leadership better connected and more nationally aware and active scientists, such as Princeton physicist Karl T. Compton (who became president of MIT in 1930) and Vannevar Bush (who became MIT's dean of engineering in 1932). These men and others soon were equally frustrated by the executive committee's inability to get anything done. One practical difficulty stemmed from the Association's financial state. But the committee's chairman emphasized trivia while ignoring significant policy issues, and Compton complained that "Cattell wastes a lot of time." Others agreed that "we never discuss anything of importance" and "never consider the long-range affairs of the Association." A committee on organization in 1934 emphasized the "confusion and ineffectiveness" caused whenever others "short circuited" an organization's chief executive. But this observation had no impact on the conduct of AAAS business. Compton and Bush, though nominally still members of AAAS committees, started missing as many meetings as their predecessors and did little at those they did attend. In 1932, as Cattell argued that the AAAS should urge Compton to accept the AAAS permanent secretaryship, Wilson wrote that nobody with Compton's qualifications "would work with Cattell. None ever has. None ever will."[109]

Though Ward had successfully led other organizations, at the AAAS he was as hard as Cattell to work with, and he alienated the chairmen of three successive AAAS local arrangements committees. The executive committee thus maneuvered him from his position, and in 1936, it named astronomer Forest R. Moulton as AAAS permanent secretary. Though Moulton's earlier career (like Ward's) combined scientific and administrative successes, his accomplishments outshone his predecessor's. His development (with his University of Chicago colleague, geologist Thomas C. Chamberlin) of the well-respected planetesimal hypothesis on the origins of planets had led to a stellar scientific reputation and election to the National Academy. In the same way, after his early retirement from the University of Chicago, he served as financial director of the city's Utilities Power and Light Corporation and then as director of concessions for the 1933 Century of Progress World's Fair. Moulton soon provided the association with the kind of leadership it had never previously experienced.[110]

The Continued Impact of AAAS Meetings

In the 1930s, the AAAS expanded its program to include regular semi-annual meetings both in December (as it had since 1902) and June (going beyond the few that had been held in the 1920s). Though many of these proved successful, not all attracted the thousands of presentations that the largest meetings of the 1920s had. After all, the Depression limited members' ability to travel, and meetings held in smaller cities like Syracuse and Richmond drew less well than others. But even these often produced excitement. In Atlantic City, in December 1932, Arthur H. Compton and Robert A. Millikan, both Nobel laureates in physics, debated the nature of cosmic rays so vigorously that after their session Millikan refused to shake Compton's hand.[111] The largest meetings attracted many affiliated societies, and the June 1933 meeting held at Chicago's Century of Progress World's Fair was especially noteworthy, as the Fair's organizers had the AAAS invite many leading European scientists, whose expenses they paid. The American Physical Society (which barely acknowledged AAAS support for its guests, such as past and future Nobel laureates Niels Bohr, John Cockcroft, and Enrico Fermi) advertised its sessions as "perhaps the most important" in its history. A symposium on "Nuclear Disintegration" attracted much attention and featured presentations by Cockcroft, Merle Tuve (who reported on the research the AAAS had recognized in 1930), and Ernest O. Lawrence of Berkeley, who described his cyclotron. The *New York Times* gave his report front-page coverage, and *Time*'s reporter praised the clarity of his paper, especially in comparison to his colleagues'.[112]

The AAAS Press Service made sure that such press and radio coverage became a major feature of AAAS meetings through the decade. Newspapers featured Millikan's failure to shake Compton's hand in 1932, and (as noted earlier) *Time*, *Newsweek*, the *Literary Digest*, and *Life* often reported on several sessions at each meeting. So too did the religious press, and magazines like *Catholic World, Christian Century,* and *Commonweal* typically featured two kinds of presentations: those whose authors seemed "quite sure of themselves" and those that stressed "how little scientists know."[113] Late in the decade, as Moulton assumed oversight for AAAS publicity efforts, he tried to expand public relations activities to include outreach programs designed to increase the public understanding of science. By 1938, the AAAS was helping NBC produce such radio programs as *Science Everywhere, Science on the March,* and *Science in the News.* But financial concerns always limited AAAS participation in such ventures; for example, 1938 plans to produce high-school syllabi to accompany *Science Everywhere* broadcasts never materialized. Ties with print media remained especially strong, however, and by the end of the decade, AAAS officers recognized a mutually profitable relationship with the National Association of Science Writers, with which they later developed the AAAS-Westinghouse Science Writing Awards (today the AAAS Science Journalism Awards).[114]

In promoting exhibits at its meetings, however, the association only slowly reconciled its conflicting goals of illustrating current research and making money from commercial exhibits. In 1931, physicist F. C. Brown—who had previously overseen research exhibits—became overall exhibits manager, and in 1935, the AAAS adopted a "general policy" that banned exhibits that could not "be clearly defended for their [scientific] value." Distinguishing between the defensible and indefensible remained difficult, however, and during the Depression, neither the association nor Brown (who worked on commission) ever wanted to exclude any potential exhibitors. Early in 1937, *Life* magazine illustrated its long report on the December 1936 meeting with photographs of mice and models of telescopes exhibited by researchers in Atlantic City. By 1940, plans for the annual exhibit went beyond anything previously produced and emphasized a mix of commercial and academic exhibits, a science library, displays of science teaching materials, and provisions for attracting public attendance. Its success laid the groundwork for the highly successful postwar exhibits.[115]

Old and New Publications

Science changed little through the 1930s, continuing to accept without review all manuscripts submitted by any well-known scientist. This practice led at times to trivial articles by Nobel Prize winners long past their prime, or by their students, though in 1936 Albert Einstein published in *Science* his concept of gravitational lensing. Despite this achievement, Cattell received many complaints about specific articles that his old-boy network brought to *Science*. He always defended them by citing the stature of the scientist who submitted the item or recommended publication. Critics also often compared *Science* to *Nature* which, they claimed, omitted trivial articles and ceremonial addresses, often published work of greater significance, and did not flood each issue with pages of advertising. Cattell's standard response compared the twelve-dollar annual subscription fee for *Nature* (which had no ties with the British Association for the Advancement of Science) with the three dollars he received for each subscription to *Science*.[116]

By the late 1930s, however, as criticisms continued to mount, he arranged to have many biological submissions reviewed by his eldest son, McKeen, himself an eminent physiologist. Other observers claimed that the reports in *Science*'s "Notes and News" columns meant less to American scientists of the 1930s than they had to the much smaller community of the 1890s. As Cattell ignored these criticisms, some AAAS members began complaining directly to Ward and then to Moulton. By 1940, Moulton had developed a standard response that "agreed fully" with the writer "regarding the high character of *Nature*," noted that Cattell retained "absolute control" of *Science*, and cited Cattell's "advanced age" to justify the AAAS's decision not to challenge this control.[117]

But even in acknowledging criticism, Moulton always emphasized that "on the whole, *Science* has been a very successful journal." Despite its problems American scientists in the 1930s and 1940s still found *Science* indispensable. The AAAS thus both looked forward to assuming control of *Science* and feared its financial consequences; despite occasional long discussions on "the future of *Science*," the executive committee always concluded that "no action was necessary" at any given time.[118] In 1936, however, when Cattell offered the association *The Scientific Monthly* on terms similar to their agreement about *Science*, further action became necessary. The executive committee at first agreed to recommend acceptance. But three days later, after it had formally named Moulton as permanent secretary, it reconsidered its vote and looked the gift horse in the mouth. Moulton's business experience led him to realize that the AAAS had no income and expense estimates, much less an audited report, and thus could not determine the publications' cost of operation nor project the proposed annuity eventually due Josephine Owen Cattell. He convinced his colleagues not to act without further information, and through the months that followed, he, Cattell, and Livingston (who had become the AAAS's general secretary) bargained intensely.[119] Moulton's regular reports to his colleagues on these negotiations scared them, and in

April 1938 Edwin G. Conklin, former president of Princeton, scrawled next to the minutes containing one such report, "This is not to the advantage of AAAS." The agreement Cattell and Moulton finally signed late in 1938 served largely, however, to protect Cattell's interests, even as it formally transferred ownership of *Science* and the monthly to the association. It called for Cattell to retain full editorial control of both journals until his "decease . . . or voluntary relinquishment," specified AAAS payments for subscriptions even after such relinquishment, and called for increases in the annuity due Cattell's widow or her heirs as cost of living rose.[120]

Early in the 1930s, the association introduced another new mechanism to advance scientific research by beginning to publish a series of short books comprising records of major symposia at AAAS meetings or reports of its committees. In 1934, it launched a more formal series of Occasional Publications, officially published as Supplements to *Science* and printed by Cattell's Science Press. The first, "The Protection by Patents of Scientific Discoveries," had a major impact on universities by urging the creation of institutional research foundations to enable universities to profit from their faculty's research achievements.[121] Despite this success, financial pressures forced the AAAS to reject some proposed Occasional Publications, though it readily issued others when sponsorship became available. For example, the publication of a 1935 symposium on "The Scientific Aspects of Flood Control" became possible when the Rural Electrification Administration agreed to purchase a thousand copies "for private distribution." Other symposia published in the 1930s included several on medical topics, such as "Cancer," "Tuberculosis and Leprosy," and "Syphilis." By the end of the decade, this series had been renamed "AAAS Selected Symposia."[122]

Advocating for Science in Response to the Depression and Fascism
The Depression and the rise of fascism led to unprecedented calls for social action on the part of most American institutions, and as early as 1931 President Herbert Hoover's Organization on Unemployment Relief asked for AAAS "cooperation." The executive committee replied, however, that "there appeared to be nothing which the Association could do." But after Franklin Roosevelt's election AAAS officials began to see in the New Deal many opportunities to win support for science. The new initiatives of the mid-1930s began a process of redefinition, through which AAAS leaders gradually came to see "the advancement of science" as necessarily requiring engagement with broader social issues, not simply coordination across scientific disciplines or the defense of the scientific community and its activities. This redefinition

did not proceed smoothly, of course, and it initially focused on issues most immediately relevant to individual scientists, such as government support for their work. In June 1933, for example, *Science* began publishing a series of notes on "Curtailment of Scientific Work Under the Government" and, especially, on "Proposed Grants for Scientific Research" to be supported by the New Deal. In December, the council voted to urge "that provision be made for adequate scientific and technical cooperation in the planning and execution of [recovery and reconstruction] projects." Within a few months, it even began to seek direct ways "for aiding unemployed scientists through Government funds for emergency relief."[123] In 1934, Ward planned a "Project for Scientific Aid for Public Works" that sought payment of $2.6 million for six months of work, and the AAAS later endorsed a bill that called for "fellowships for unemployed scientists." At about the same time, the AAAS joined other groups in supporting Karl Compton's call for a $16 million "Recovery Program for Scientific Progress." But none of these proposals went anywhere, and by the end of the year, the council resolved "that aggressive governmental support of scientific work is essential to any sound program of building for the future national welfare."[124]

The new attraction to social issues as part of the advancement of science became further apparent as the AAAS struggled to find appropriate responses to the rising tensions in Europe, especially the effects of fascism on science. The AAAS first took notice of these in January 1932, as *Science* reported American protests against Mussolini's dismissal of Italian professors who refused to swear allegiance to his regime. After Hitler's rise to power early in 1933, *Science* reprinted from other journals denunciations of Nazi dismissals of Jewish professors. Later that year, *Science* published reports of the British Academic Assistance Council, the Committee of Dutch Professors on Behalf of German Jewish Students and Graduates, and the New York–based Emergency Committee in Aid of Displaced German Scholars. In December 1933, the AAAS council adopted the "Declaration of Intellectual Freedom" that condemned (without mentioning the situation in Germany) "persistent and threatening inroads upon intellectual freedom which have been made in recent times in many parts of the world" as a "major crime against civilization itself." But through the mid-1930s, the association itself officially responded hardly at all to events in Europe or east Asia.[125]

In Britain, meanwhile, a significant number of scientists began conflating concerns about fascism and the impact of the Depression in what soon became known as the Social Relations of Science movement. The BAAS featured sessions on "the impact of science on society" and "the scientist's responsibility for the social uses of science," and Sir Richard Gregory used his editorship of *Nature* to advocate these interests.[126] And though similar concerns echoed in parts of the American scientific community—some even called for a "science holiday" or "research moratorium" to give the larger society time to consider these issues—they remained largely unrecognized among AAAS influentials. After all, many of these influentials had served on the council and the committee on policy in 1919, when the association had voted to avoid dealing with "political questions on which public opinion is divided." In 1935 *Science* even broke tradition by refusing to publish a Society for the Promotion of Engineering Education presidential address, apparently because it called for engineers to examine the social and economic effects of their work.[127]

The following year, however, the AAAS began to move away from this position. In June 1936, AAAS president Edwin G. Conklin attended a BAAS meeting dedicated to "Science and Social Welfare," which passed resolutions calling for the AAAS to join the BAAS in promulgating a "Magna Charta for Science." On his return to America, Conklin convinced his AAAS colleagues to sponsor "a series of conferences" to be held at semiannual meetings "on the general subject of science and civilization."[128] Soon thereafter more politically radical AAAS members called for an "avowedly progressive" American Association of Scientific Workers that looked "beyond liberal reform" to advocate radical responses to Fascism and the lingering effects of the Depression. Once organized, the AASW met regularly with the AAAS, and it formally became a AAAS affiliate in January 1943.[129]

More politically mainstream initiatives proved much more influential. Most notably, in October 1937, Moulton presented to the executive committee a series of reflections on the association and its goals that looked more fully to the future than any report this committee had ever discussed. This document bluntly asked, "Is there . . . any important work that the Association should do and can better than any organization?" and concluded that "if there is not, having served its day and generation well, it should now fold up its tent and silently steal away." Not surprisingly, Moulton did believe the AAAS had a major role to play in the distressed world of the 1930s through public outreach programs (largely embodied in attempts to increase the association's "lay membership"), publications (with *Science* for its professional members and a revised *Scientific Monthly* designed to serve other members), and especially education. But this executive committee document called most immediately for the AAAS to defend science against those who attacked its impact on society and called for a research moratorium and, especially, against the "arbitrary controls" on science then emerging "in several great countries."[130]

In outlining a defense, Moulton cited a recent *New York Times* editorial (reprinted in *Science*) that praised British proposals (which grew from previous calls for a "Magna Charta" for science) for the BAAS and AAAS to form the "nucleus of a democratic 'World Association of Science' concerned with the solution of international social problems."[131] That December, the AAAS council voted both to "make as one of its objectives an examination of the profound effects of science upon society" and to invite other organizations, especially the BAAS, "to cooperate . . . in promoting peace among nations and intellectual freedom in order that science may continue to advance and to spread more abundantly its benefits to all mankind." These resolutions received much press attention, as did many related activities at the December 1937 meeting.[132] These included Conklin's address as retiring president on "Science and Ethics," symposia on the "Social Implications of Modern Science," and even Stuart Gager's Botanical Society of America presidential address, which addressed "Pandemic Botany." Moulton's report on the meeting in *Science* described it "as the beginning of a new era in the association" and emphasized "the increasing sense of responsibility of scientists to society" that pervaded many of its sessions. Though Moulton may have exaggerated, and not all AAAS meetings in the years that followed addressed "science and society" issues as fully as this one, the association abandoned its 1919 decision to avoid such concerns. In 1937, they clearly had become a major part of the AAAS agenda, where they remain today.[133]

International programs proved less successful in the late 1930s. The AAAS council readily resolved to cooperate "in principle" with the BAAS, but financial concerns on both sides limited action. Moulton also feared entangling the AAAS with the radical Marxists who played major roles in BAAS affairs. He thus emphasized that past AAAS resolutions explicitly excluded any "criticism of governments or of social orders." Such statements cooled British ardor, and in July 1938, the two associations agreed merely to alternate sponsoring sessions at each other's meetings. Some months later, Moulton wrote in *Science* that the spirit of science—"tolerance, cooperation, and the rule of reason"—had just saved the world from a "plunge into the abyss of another world war" through the Munich agreement, which gave Hitler control of part of Czechoslovakia.[134] In December, however, the *New York Times* introduced its coverage of the association's meeting with the headline "Scientists Gird to Rescue World from Misuse of Man's Inventions." The first fruit of AAAS-BAAS cooperation reflected such concerns, as *Nature*'s Richard Gregory spoke on "Science, Religion, and Social Ethics." Other politically charged sessions included a meeting of the section on anthropology, at which all who attended signed statements denouncing Nazi racial theory, and Percy W. Bridgman's presidential address before the American Association of Physics Teachers, which anticipated his "Physicist's Manifesto" of the following January, when he closed his laboratory to citizens of totalitarian states.[135]

Education and Outreach Programs

Despite Moulton's October 1937 call for the AAAS to emphasize science education programs, this interest remained peripheral to most AAAS influentials through the decade. The AAAS's committee on the place of science in education did continue earlier activities. These included support for junior academies of science, the circulation of popular reading lists, educational exhibits and sessions for science teachers at many semiannual meetings, and the radio programs noted earlier. And though longtime general secretary Otis W. Caldwell and his colleagues devoted much time to them, none ever proved as significant as they had hoped. Limited financial resources restricted their actions and stifled even well-planned programs.[136] For example, in 1934, as chair of a committee on adult education, Harlow Shapley built on his experience with educational broadcasting in Boston to sketch plans for a "national radio university" through which listeners could earn college credit and "programs for all grades of intellectual maturities." Realizing that his committee could not go further without AAAS support, and he knew that such support was not forthcoming, he expressed regret that "it will not be possible to do anything justifying the expenditure of much time and energy of five fairly valuable members of the scientific community."[137] Three years later, Moulton's document looking toward the future envisioned "service to the youth of America" as "the easiest extension of the work of the Association" and projected a category of junior members, with their own weekly *Junior Association Magazine*. His executive committee colleagues ignored these notions, which went nowhere.

One AAAS science education program proved highly successful in the 1930s and afterward, however, though it reflected more inward-looking concerns than did Shapley's ideas. In 1937, Neil E. Gordon, secretary of the section on chemistry, sought AAAS sponsorship for the summer research conferences he had directed since 1931. These had originated at Johns Hopkins and at first emphasized tutorials on research methods. By 1937, they had migrated to the Gibson Island Club, a private resort on the Chesapeake Bay, and their format had evolved significantly. Daily sessions at each week-long conference opened with informal presentations of "frontier problems" by leaders in particular fields. Gordon kept afternoons free for recreation and—both by carefully selecting participants for their ability to address the work at hand and by banning any record of proceedings—he promoted free-form evening discussions that typically continued long into the night. The conferences had established a fine reputation in the early 1930s, as such eminent scientists as Leo Baekeland, Irving Langmuir, E. V. McCollum, W. A. Noyes, and Harold Urey led sessions on, respectively, "Synthetic Resins," "Surface Phenomena," "Vitamins," "Photochemistry," and "Heavy Hydrogen." As these titles suggest, by the mid-1930s Gordon had begun to look beyond chemistry, and he ensured his program's success by attracting support from over thirty industrial firms. In 1937, the AAAS executive committee voted to sponsor the conferences, "subject to the condition that the sessions be held without expense to the Association." The AAAS continued its sponsorship on these terms into the 1940s, and even helped secure a mortgage on the Gibson Island property. The conferences continued throughout the war years even when the AAAS had to cancel its own meetings; in 1947, they relocated to several sites in New Hampshire, and the following year they became the Gordon Research Conferences. Their meetings and impact continue today, as the conferences remain officially a "participating organizaton" of the AAAS, and *Science* publishes their schedule twice yearly.[138]

Another form of outreach promoted by the AAAS in the 1930s involved the creation of local branches. As early as 1913, it had authorized communities to establish such branches, and though one or two appeared by 1915, the association's expansion interests soon focused on the establishment of broader groups like the Pacific Division, and the local branches soon died out. In 1934, however, the initiative reemerged in an attempt to reinforce member loyalty during the Depression. Within a year, branches appeared in Lancaster, Pennsylvania; Mobile, Alabama; Phoenix, Arizona; and southern Rhode Island, and the AAAS recognized the New York Institute of Science (in Westchester County, just north of New York City) and the Southern Florida Science Association as local branches. But only the Lancaster branch survived, as Cattell's son Jaques (who managed his father's Science Press Printing Company) used it as a base to promote all sorts of cultural events. For years its membership surpassed 1,000.[139]

Responding to the War and Preparing for the Post-War World

World War II mobilized American society as never before. Within the scientific community, many American scientists devoted eighty hours or more each week to the war effort, mostly through federal agencies or national laboratories. The AAAS itself focused less attention than many other organizations on the war itself. As wartime restrictions limited travel, it canceled its 1942 and 1943 meetings. As early as 1942, however, some AAAS members began making and debating plans for science in the postwar period, and in 1943, many articles in *Science* called for support for the Science Mobilization Bill introduced by Senator Harley M. Kilgore of West Virginia. This proposal helped lead, eventually, to the National Science Foundation, the creation of which the AAAS actively supported in the late 1940s.[140] But through the decade's first half, internal matters and planning for the association's postwar future dominated the attention of most scientists who could devote any time at all to AAAS affairs.

Most concerns about the association and its future reflected the stifling control that Cattell continue to wield, about which many AAAS members had long complained. From the moment he took office, Moulton began to challenge this dominance as none of his predecessors had. To be sure, the 1938 negotiations that led to the contract on the future of *Science* and the *Scientific Monthly* served (despite Moulton's best efforts) Cattell's interests more than the association's. But more serious challenges emerged in 1940, when Albert F. Blakeslee, a geneticist at Carnegie Institution of Washington's Cold Spring Harbor Laboratory, became AAAS president. By the end of his term, after working closely with Cattell and Moulton and failing to get the financial reports on the journals that his predecessors had sought, Blakeslee broke precedent and prepared a long "Report of President to Council" that called for "a careful and sympathetic study of the problems of the journals."[141]

Despite these conciliatory words, Cattell knew that Moulton and Blakeslee hoped to reduce the association's financial commitment to his estate. He also believed that Moulton (whom he distrusted) wanted to become *Science*'s next editor, and he told friends that he hoped that would never happen. Some responded (as Harlow Shapley did) by asking "why should not a man of 81 be entitled to whatever [he wants?]." In the months that followed, the "careful and sympathetic study" evolved into plans to challenge the contract on the grounds that Cattell (as owner of *Science* and as chairman of the AAAS executive committee) was a party on both sides. Breaking the contract proved impossible, however, though in April 1941 pressure from others finally led Cattell to resign as chairman of the AAAS executive committee, a position he had held since 1925.[142] The AAAS soon asked if Josephine Owen Cattell would accept a lump-sum payment, rather than the annuity the contract called for. The Cattells actively opposed the suggestion and after the attack on Pearl Harbor few American scientists had time to pursue the matter. In March 1942, however, Moulton began issuing a monthly eight-page *AAAS Bulletin* reporting news of the association. (Such news remained limited during the war.) Tensions between Cattell and Moulton gradually increased and become more personal, and soon they became even worse.[143]

Moulton had meanwhile begun to edit the *Scientific Monthly*, while Cattell's son Ware served as associate editor. This arrangement continued until early 1943, when Moulton's other responsibilities forced him to give up the office. Ware then became editor, a position he had coveted for years, but he brought to his editorship eccentricities that his father had tolerated but that Moulton did not, and when Moulton tried to rein him in, the younger Cattell complained to his father. Ware also spied on Moulton for his father and shirked his duties even after he became editor. In July 1943, after months of Ware's negligence, Moulton fired him. The younger Cattell soon sued the AAAS and collected much of his unpaid salary. For the rest of the year, Cattell listed his son on *Science*'s masthead as assistant editor. The AAAS's two most powerful figures remained estranged until Cattell's death in January 1944.[144]

The control of *Science* then passed immediately to the AAAS, and though they had long expected this transfer, the association's leaders were surprised by its timing. They rushed to ensure that Josephine Owen Cattell and her son Jaques would continue as editors (though without published recognition) and, after receiving audited financial reports on *Science*'s income and expenses, again tried to renegotiate the terms of the association's contract with the Cattells. The family would not even consider doing so. The association finally decided to bite the bullet, and in 1945, it learned that the annuity it owed the Cattells (before inflation) totaled over $165,000. By 1954, when all the payments were completed, the association had paid about $270,000 to the Cattell family.[145]

Providing for *Science*'s editorial future proved even harder. After considering several eminent scientists, in August 1944, the executive committee named Charles S. Stephenson, a retired naval physician, as *Science*'s editor as of January 1945. Stephenson's first chore involved planning to take over all of the editorial, business, and production duties that the Cattells had performed since 1895 without charge to the association. The proposal he prepared that fall called for large offices, a staff of eleven, and a salary budget of $32,000. This plan gave the members of the executive committee their first notion of what it cost to edit and produce *Science* and, since they knew the association could not afford what Stephenson had proposed, they told him that they had to accept his resignation. He soon sued for breach of contract, and the AAAS later settled his suit out of court.[146] In January 1945, *Science* announced that Josephine Owen Cattell and Jaques Cattell would serve as editors for the year, subject to the oversight of a policy committee on *Science*. As its members tried to micromanage the journal, tensions between the Cattells and the AAAS continued high, even as the editors formally proposed renewal of their appointment. But the Association never considered this proposal seriously; after all, Josephine Owen Cattell was then seventy-four years old, and neither she nor her son had scientific credentials. In October 1945, the executive committee chose Willard Valentine to edit *Science*, beginning the following January.[147]

Meanwhile, as both the association and the Smithsonian Institution planned to take on new responsibilities in the postwar world, the AAAS knew that it would soon have to move its headquarters. During the war, office space in Washington grew scarcer by the day, and in August 1944, the executive committee authorized Moulton to swallow the association's pride and formally request "space in the National Academy Building for the offices and the publications of the Association." Nothing came of this request, however, or from others that soon followed. The association's move to its first permanent headquarters on Massachusetts Avenue in Washington took place only after the war's end.[148]

More significantly, through the mid-1940s, the executive committee comprised a cohort of strong figures, including Livingston (who succeeded Cattell as the executive committee's chair), chemist Roger Adams, physiologist Anton J. Carlson, and Harvard president and chemist James B. Conant. After 1943, when victory seemed inevitable and they could devote themselves to AAAS affairs, these men worked with (and at times restrained) Moulton and drew lessons from the association's difficulties during the previous several years. In 1945, for example, as the association settled the suits brought by Ware Cattell and Charles Stephenson, its then president, Charles F. Kettering of General Motors, an eminent engineer with even wider business experience than Moulton, reminded his colleagues that "all business relations . . . in which there could be a liability" should be reviewed by counsel. They also realized that the association required greater continuity in governance than that provided by its constitution of 1919. They proposed a constitution that replaced one-year presidential terms with three-year terms for leaders who served, successively, as president-elect, president, and immediate past president. Though still elected by the council, those filling these offices also held seats on the executive committee, a procedure that greatly enhanced the association's governance both by providing continuity and by involving additional new people in the AAAS leadership.[149]

Most significant, however, as the association and its leaders (especially Harvard geologist Kirtley F. Mather) considered the impact of the war on their world in light of the Social Relations of Science movement that had occupied so much of their attention in the late 1930s, they looked more than ever before to expanding the association's purposes. In its first hundred years, the AAAS had successfully helped define the scientific community, arguing especially for the importance of encompassing the full range of disciplines and practices inherent in a democratic society. Through its annual meetings and through *Science*, it had provided forums through which this community expressed its identity and—at a more practical level—exchanged the ideas that comprise the scientific endeavor. Council resolutions and the vigor of individual AAAS leaders had effectively advocated for science in a variety of settings. But in the early 1940s, as these leaders contemplated postwar needs, they proposed a new constitution, rewording objectives that had not been altered since 1848. Even more important, they added two significant clauses to these objectives that reflected new visions both about the potential of the overarching interdisciplinary perspective that the AAAS offered, and about what the phrase "advancement of science" meant. These new clauses called for the association "to improve the effectiveness of science in the promotion of human welfare, and to increase public understanding and appreciation of the importance and promise of the methods of science in human progress."[150] As the chapter that follows demonstrates, these clauses set the agenda for the association's most recent fifty years.

CHAPTER 3

Shifting Science from People to Programs

AAAS in the Postwar Years

BRUCE V. LEWENSTEIN

$\mathcal{A}$s World War II ended, both the United States and American science took on new roles worldwide. For the country, the end of the war marked its emergence as a world power. For science, the war marked the emergence of American research from the shadow of European forebears. Most dramatic, of course, were the products of science: radar, jet engines, penicillin—and the atomic bomb. While scientists and historians might quibble about the relative contribution of "science" and "technology" to these products, they were widely seen both as dependent on science and as significant contributors to the successful winning of the war.

Redefining "Advancement"

The products of science were tied to a host of new contexts and institutions for science in the years during and immediately after the war, years that saw a new public enthusiasm for science, replacing the anger at science sometimes common during the Depression years. The immigration of leading physicists, mathematicians, and others from Europe in the 1930s, the dramatic growth of federal funding for research, the rapid expansion of institutions like the National Institutes of Health, the emergence of the civilian National Science Foundation out of the research traditions begun during and after the war by the Office of Scientific Research and Development (OSRD) and the Office of Naval Research—all these contributed to the creation of a new relationship between science and other social institutions.

In that new context, the nature of the AAAS had to change. From its origins through the 1930s, the AAAS had described itself largely as an organization of people, a group that "through its meetings and through its publications . . . promotes intercourse and co-operation and the feeling of fellowship among men and women of science and all who are interested in the progress of knowledge and education."[1] But as science grew and the differences among individual researchers in diverse fields became so great that they shared little fellowship, the AAAS found a need to transform itself into an organization independent of the people who nominally constituted it, an organization with its own programs, supported by the dues and sometimes actions of its members, but ultimately run by a professional staff committed to the advancement of science.

New Visions

As the war ended, the AAAS was the leading interdisciplinary scientific society in the United States. Its twenty-eight thousand members were well placed to claim the mantle of meeting the nation's needs in scientific areas, but the organization was facing an era of change as specialized societies took over many of its functions. Moreover, the AAAS had cut back its activities substantially during the war years, and now it was rushing to catch up. First, it expanded its membership; from an average of about a thousand new members each year in the late 1930s and through the war, the membership shot up by five thousand in 1947 and by ten thousand the following year (partly as the result of a special membership drive for the association's hundredth year). Although the pace slowed during the 1950s, by the mid-1960s, the AAAS would be adding ten thousand new members per year for years on end.[2]

Immediately after the war, the AAAS updated its constitution. The objectives of the association had remained essentially unchanged since its creation in 1848, and focused on the needs of a scientific community seeking to define itself in a vast country with uncertain communications among its parts. The AAAS's goals had been "to promote intercourse among those who are cultivating science in different parts of America, to coöperate with other scientific societies and institutions, to give a stronger and more general and systematic direction to scientific research, and to procure for the labors of scientific men increased facilities and a wider usefulness."[3] As the first century of the AAAS came to a close, however, the scientific community was clearly established. Perhaps feeling more secure, the association sought a new ecological niche. With the transfer of AAAS power from James McKeen Cattell and the old guard, the association's goals changed. Although the British Social Relations of Science Movement never took strong hold in the United States, many of the leading AAAS figures in the 1940s were influenced by it, including Administrative Secretary F. R. Moulton. Although they recognized the new public support for science, the new AAAS leaders also believed they needed to demonstrate responsibility and awareness of social issues to justify that support. These leaders urged the scientific community to redirect its attention outward, both to serve society and to ensure that science and scientists achieved a more secure place in the wider society. Thus the new AAAS constitution, developed during the years of World War II and adopted in 1946, set new objectives: "To further the work of scientists, to facilitate cooperation among them, to improve the effectiveness of science in the promotion of human welfare, and to increase public understanding and appreciation of the importance and promise of the methods of science in human progress."[4] These new objectives set the tone for the next fifty years of AAAS history, in which the association would continue to provide for interchange among the individual sciences (at both the national and, increasingly, international levels) without necessarily being the locale for those interchanges; help direct science toward serving public needs; help shape science education; and promote the value of science to government and public audiences.

The constitutional changes were deliberate and followed on the changed attitudes towards issues of science and society that had emerged in the 1930s. As Moulton said in his summary of the new constitution, "From 1848 until 1946 scientists, according to the constitution, were looking inward; now they are looking outward. In these earlier years they were primarily interested in science and themselves; now their vision encompasses all humanity. Formerly they assumed that the advancement of science depends upon scientists alone; now they realize that it rests also upon the understanding and appreciation of the public. Science has been transforming the physical environment of human beings; now it promises to provide patterns of thinking and ways of living."[5] Leaders of the postwar AAAS knew that their concerns about promoting science could not be addressed without considering the relationship between science and the wider public.

The most immediate wider public was the federal government, and the most important issue facing the scientific community was the proposal for a new National Science Foundation. Competing versions of legislation to create the NSF split the leaders of the community; some believed that only scientists should govern the NSF, while others accepted that a public agency should be overseen by a presidential appointment. A committee created by the AAAS in 1945 to provide advice on the various competing legislative proposals evolved into an Inter-Society Committee for a National Science Foundation, which provided a forum for much of the discussion among scientists. Various proposals were shot down in Congress or at the White House. A 1950 meeting of three officers of the Inter-Society Committee with two members of the U.S. Bureau of the Budget led to the crucial compromise (in which the NSF is run by both the presidentially appointed director and the National Science Board, consisting of prominent scientists); of the five people at that meeting, one (Harlow Shapley) was a former AAAS president and two (Dael Wolfle of the Inter-Society Committee and William D. Carey of the Bureau of the Budget) were future AAAS executive officers.[6]

In addition to examining the political arena, the new generation of AAAS leaders also began to direct their attention to the association itself. In the first five years after the war, the AAAS tried several times to reconsider its activities in light of its newly stated goals. Some of those changes were minor, such as increasing use of the *AAAS Bulletin* as a way to reach members. (The *Bulletin* had been introduced by Moulton in 1942 as a means of communicating directly with members, without having to go through the aging Cattell at *Science*.) But others were more substantive. Early in the 1940s, the AAAS had once more launched a campaign to promote the need for more interdisciplinary links. "As science becomes more and more specialized," according to an article in the launch issue of the *AAAS Bulletin* (presumably written by Moulton), "it becomes increasingly important to maintain interconnections among its various fields." The symposia and other meeting sessions that were jointly organized by several AAAS sections or affiliated societies served that purpose, but regular meetings were not enough, the editors said. The AAAS "is ideally constituted to furnish opportunities for exploring the interrelationships of science and society."[7]

The most dramatic attempt to implement this new vision was the 1948 centennial meeting of the association. Traditionally, the meeting was a time for technical papers and specialized science, with many of the smaller scientific societies that were AAAS affiliates providing the detailed programs for the AAAS. But, at the insistence of 1947 AAAS president Harlow Shapley (a Harvard astronomer), the 1948 meeting in Washington, D.C., was revamped to include only symposia that covered broad synthetic areas.[8] U.S. president Harry Truman delivered the keynote address, endorsing the proposed National

Science Foundation and committing himself to open communication in science—a direct challenge to the increasing attacks on scientists coming from the House Committee on Un-American Activities. Individual sessions on genes and cytoplasm, energy sources, and waves and rhythms filled the mornings; open houses and tours of the many scientific agencies and organizations in Washington took over the afternoons; and public lectures filled the evenings. A media success, the meeting was a scientific disaster, with attendance dropping from nearly 5,000 the year before in Chicago to only 2,700.

Moreover, this initial attempt to redirect the association backfired. Many of the smaller societies—especially the biological societies—had long been concerned about their role in AAAS. The 1948 meeting pushed them to decide that a separate forum would be a better place for their own technical sessions, so they created the new American Institute of Biological Societies. Since the biologists had long been the major supporters of the AAAS annual meeting, their departure created a crisis of sorts. Other pressures for change came from a new generation of AAAS leaders, unshackled from prewar allegiances, who deeply believed that the AAAS had to make its programs more relevant to a world in which science could no longer insulate itself from the political, economic, and social currents around it.[9] One of these new leaders was Howard A. Meyerhoff, a geologist who replaced F. R. Moulton when he retired as administrative secretary in 1949. Meyerhoff had worked for AAAS in the past, and had a clear interest in policy issues—he had followed the legislative peregrinations of the National Science Foundation (which took nearly five years to establish) in detail, and had a strong interest in problems of "scientific manpower" (the need to meet the demand for more scientists in the postwar world). Meyerhoff argued that future meetings should include some combination of specialized and generalized symposia.[10]

Most important among the new postwar leaders, however, was mathematician Warren Weaver, director of the Natural Sciences Division of the Rockefeller Foundation, who joined the AAAS executive committee in 1949. Weaver represented the rapid movement in science away from lone investigators and toward managed, institutionalized research.[11] Weaver had become interested in reaching beyond the scientific community at the end of World War II, when he coordinated a series of speeches made by scientists and broadcast during the intermissions of New York Philharmonic radio broadcasts. Weaver turned to the AAAS as a vehicle for helping scientists reach out to the public.[12]

Despite Weaver's urgings, the AAAS did little to build new programs until the early 1950s. Most of the energies of its leaders went into bringing its flagship journal, *Science*, to a new maturity away from the Cattell family, which had turned the journal over to the AAAS shortly after Cattell's death in 1944.[13] But the concern among AAAS members about the association's role reached a breaking point in 1951. That year, after the American Institute of Biological Societies once again decided not to hold its annual meeting in conjunction with the AAAS, executive committee member Paul Klopsteg wrote that "the conclusion may be drawn that . . . the function of A.A.A.S. in serving scientific societies as a forum for exchange of ideas no longer exists."[14] In June 1951, former AAAS president Anton J. Carlson called for a formal reevaluation of AAAS goals. Carlson, a prominent physiologist from the University of Chicago, joined two colleagues in a letter to the AAAS executive committee that challenged the AAAS to "find another function or die." Though Carlson and his colleagues were concerned by internal problems at AAAS, they were also stimulated by events outside the scientific community; they were especially disturbed by challenges against the scientific community developing as anticommunism evolved into the witch-hunting accusations that would become the McCarthy era of the early 1950s, leading political liberals of all kinds to be tarnished as Communists or sympathizers. Carlson and his colleagues were motivated, they said, by the "dire threats to the advancement of science" that grew "out of fears heightened by the present world political struggle."[15]

Carlson and his colleagues were well known to be political liberals, but they were not alone in their response to political problems. Throughout the 1950s, many scientists worried about the effects of rabid anticommunism, which then engulfed much of public political debate. The issue was expressed as a concern with "intellectual freedom" and "freedom of speech," concerns motivated by charges of participation in un-American activities made against prominent scientists such as Edward U. Condon, Kirtley Mather, Linus Pauling, and others.[16]

To meet the new challenge, Carlson and his colleagues argued that the AAAS should retreat from "direct nourishment of scientific effort," on the grounds that a "host of scientific bodies" now filled this role. Instead, the AAAS should undertake a "full-fledged reorientation" toward presenting science to the public. For Carlson and his colleagues, science offered not just a potential solution to the many problems facing the modern world; it offered the *only* solution. Without recognizing the irony in a democratic society of presenting themselves as having "the" solution to what they admitted to be a complex and multifaceted problem, they especially worried about what seemed to them to be totalitarian approaches used by some American politicians of the day: "The concepts of free inquiry and of nonauthoritarianism, of integrity and pragmatism—to mention a few principles of scientific activity—must be conveyed to the public or there can be neither substantial public understanding of science nor wider adoption of its effective approaches to man's problems." Carlson and his colleagues believed that AAAS offered a forum for activities that would support these goals.

Carlson's letter received a welcome response from Weaver, who was on the AAAS executive committee, and convinced the committee to schedule a special conference in September 1951 at Columbia University's Arden House retreat along the Hudson River. In a list of preliminary questions published in *Science*, Weaver tacitly assumed the older style of annual meetings to be "antiquated," and asked whether the AAAS, "as the most inclusive spokesman for science, [should] recognize as one of its main activities (perhaps the main one?) exposition and interpretation of science not only to all scientists, but even more importantly, to the general public?"[17]

The Arden House Conference

The 1951 Arden House conference defined AAAS as we know it at the close of the twentieth century. Attended by AAAS executive committee members and by a number of consultants handpicked by Weaver, the AAAS leaders examined activities directed toward scientists, activities directed toward government, and "activities relating to scientists' connection with the public and with society in general." The result, orchestrated by Weaver, was a commitment to more synthetic, interdisciplinary activities when addressing scientists, and a more active and structured presentation of scientific issues to the public. Weaver summarized the findings of the conference in a report published in the association's two magazines (*Science* and the *Scientific Monthly*) and signed by all those who attended the conference: "We have reached the stage where one over-all organization cannot effectively deal with the intensive and specialized interests of individual branches of science. The technical papers that present detailed results in chemistry, in physics, in mathematics, in zoology, etc., can more properly be presented before meetings sponsored and arranged by the appropriate professional groups." Therefore, he said, the AAAS should cease its existing pattern of "outmoded" meetings, and shift to something that would provide "a synthesizing and unifying influence." (See appendix A for full text.)

One of the best ways for AAAS to provide such an influence, Weaver wrote, would be for the AAAS "to begin to take seriously one statement of purpose that has long existed in its constitution"—the call "'*to increase public understanding and appreciation of the importance and promise of the meth-ods of science in human progress'.* . . . It is clearly recognized . . . that in our modern society it is absolutely essential that science—the results of science, the nature and importance of basic research, the methods of science, the spirit of science—be better understood by government officials, by businessmen, and indeed by all the people."

Weaver stressed that "the attendants at the Arden House Conference did not intend that this statement be viewed as a polite rephrasing which sug-gests only minor changes." Instead, it called for active reassessment and re-direction of AAAS, away from technical issues and toward more synthetic issues within science and more concern with public understanding of science in the nonscientific community.[18]

Reflecting the general concern with the political situation, Weaver stressed the government in his definition of the "public" that ought to con-cern the AAAS. At a special executive committee meeting following the Arden House conference, the AAAS agreed to cooperate with the American Chemi-cal Society, the American Geological Institute, the American Institute of Phys-ics, and other scientific societies in "a long range program of public education in matters affecting science and scientists, and a more immediate program of public education with reference to manpower in the present emergency" [the Korean War].[19]

Within the AAAS staff, enthusiasm for the possibility of revitalizing the association's programs initially ran high. During the planning for Arden House, Administrative Secretary Howard Meyerhoff actively assisted Weaver in finding consultants and others with an understanding of "public relations" and "imaginative programs" for AAAS. (Although most observers conflated "public relations" and "programs" into a single "public education" program, Meyerhoff's desire to keep them separate suggests a different perspective on what a more public orientation for AAAS ought to be. The tension between these differing visions would recur over the rest of the century in AAAS de-liberations.) And when the executive committee considered the Arden House report, Meyerhoff seconded the motion to accept the report, implying that he supported the initiatives discussed there.[20]

AAAS members also quickly expressed support for the Arden House initiatives. Within a month or two after a report on the conference appeared in *Science*, the AAAS had received nearly a hundred letters applauding the general tone of the conference. For many of those correspondents, the key issue was the inherent value of science as a way of approaching the world's problems. They presented the issue in moral terms: Only science could provide answers to many of the world's problems, and so they as scientists had a moral obligation to demonstrate that reality to nonscientists. Theirs was a missionary commitment, untempered by the inherent tension of claiming special expertise in a society committed to broad democratic decision-making. A program "passing on to government and to the general public in an authentic and easily comprehended manner, the findings of our research workers as they apply to the public welfare, is one of very great importance," wrote an engineering professor. "If we can, as scientists, become well enough organized and our *knowledge* of the great problems of the day be recognized as having more weight than the *opinions* of pressure groups and selfish politicians, we would make a very important contribution to national welfare."[21] Another scientist, University of Chicago physiologist R. W. Gerard, made it clear that scientists had absolutely no doubt about the efficacy and importance of research—especially basic research. "I take it as needing no argument," Gerard wrote, "that the importance of science to the national welfare, and the consequent imperative need for the people and their elected representatives to understand something of how science operates, makes such public education the prime duty of the AAAS."[22] This sense of national purpose captured a significant element of the emergence of science as a major institutional player in the postwar world.

Weaver supported this sentiment, though he recognized the political danger of claiming too much for science. He demurred when University of Wisconsin president E. B. Fred suggested adding to the Arden House statement a phrase indicating that scientists should "understand and help direct the impact of science upon society." Weaver agreed that scientists had a right to say how public policy should be made, but he objected to the idea that the right was solely theirs: "There are some scientists who have the idea that they are so much wiser than other persons, and that science is so superlatively and exclusively the accurate way of thinking about things, that they believe that scientists are really the ones who should tell the rest of society how to behave. I don't happen to think this."[23] Weaver's hesitation was prescient, for over the next twenty years much of the conflict over AAAS goals involved the degree of authority to be given to science.

The new attention to public understanding of science drew particular interest from many AAAS members. As historians John Burnham and Marcel LaFollette have shown, much of the popular science produced during the postwar years presented individual discoveries and facts, failing to impart a sense of science as a whole, of the scientific method, of the scientific spirit. AAAS members recognized this at the time and believed that this failure would lead to a triumph by the antirationalist forces in society—a triumph of superstition. The public would see only the magical productions of science, not the intellectual process producing them. The public did not need, supporters of this view claimed, more scientific facts. On the contrary, what it needed was some way of interpreting those facts or the confusions they engendered.[24]

Yet support for new popular science initiatives was not universal. Some scientists, concerned for the details of science, opposed the call for a reorientation of the AAAS's mission. "If we lose contact with the bedrock of specialized research," wrote one, "we . . . run the danger of shallow, vapid, unfounded generalizations. Be our integrators ever so careful, if they operate without the sharp light of detailed investigations they may fall into facile dilettantism." (This same correspondent did not reject the notion of devoting some effort to the public understanding of science, however, asking, "Must we do away with one to concentrate on the other?") Another scientist noted that while the average AAAS member "undoubtedly wishes piously that the average citizen, the government official, and the public at large had a better understanding of Science, his budget does not permit him to indulge that wish and to support an organization that makes this instruction its major objective."[25]

Still, the consensus among AAAS members seemed to be that the organization should take a leading role in popular science. Most important was the opportunity that programs in public understanding of science offered for contributing to what they saw as the health of the nation. "I have been disappointed in critical junctures recently," wrote American Council of Learned Societies executive director Charles Odegaard, "at the infrequency with which scientific leaders have cut through the heart of the matter to explain to the public that science is not only a subject matter but also a way of life embodied in the value systems of American Society." Odegaard deplored the tendency to defend science "on the gadget level," by stressing the material benefits of science. Instead, he argued, defenses should be "devoted to the spirit of free inquiry."[26]

By stressing the cultural, democratic, even patriotic goals of popular science advocates, Odegaard captured a significant theme among the intellectual community in the immediate postwar period. The broad consensus in favor of devoting major AAAS resources toward the public understanding of science reflected the postwar commitment of the newly confident scientific community to seeing its values put into widespread use. Morally certain of the effectiveness and appropriateness of their worldview, members of the scientific community hoped to see their values deeply influence the culture around them.

The Academy of Natural Sciences of Philadelphia, site of AAAS founding. *Credit: Ewell Sale Stewart Library, The Academy of Natural Sciences of Philadelphia.*

Henry Darwin Rogers, one of the founding fathers. *Credit: The MIT Museum.*

Louis Agassiz, AAAS President 1851, one of the founders. *Credit: Marine Biological Laboratory Archives.*

William Barton Rogers, AAAS President 1876, one of the founders. *Credit: The MIT Museum.*

Joseph Henry, AAAS President 1849, one of the founding fathers. *Credit: AAAS Archives; engraving by S. Hollyer.*

Alexander Dallas Bache, AAAS President 1850, one of the founders. *Credit: AAAS Archives.*

Maria Mitchell, early woman member. *Credit: the Nantucket Maria Mitchell Association.*

Dudley Observatory, Albany. *Credit: The Collection of the Albany Institute of History and Art.*

SCIENCE

A WEEKLY RECORD OF SCIENTIFIC PROGRESS.

ILLUSTRATED.

Entered in the Office of the Librarian of Congress, at Washington, D. C.

| Vol. I—No. 1. | JULY 3, 1880. | Price 10 Cents. |

First issue of *Science. Credit:* Science *magazine.*

James McKeen Cattell, Editor of Science 1895–1944. *Credit: AAAS Archives.*

Josephine Owen Cattell. *Credit: James McKeen Cattell Papers, Manuscript Division, Library of Congress.*

Frederic Ward Putnam, AAAS President 1898. *Credit: Peabody Museum, Harvard University.*

Edward C. Pickering, AAAS President 1912. *Credit: AAAS Archives.*

E. E. Just. *Credit: Marine Biological Laboratory Archives.*

L.O. Howard, AAAS President 1920. *Credit: AAAS Archives.*

Left: Burton E. Livingston, AAAS Permanent Secretary 1920–1930. *Credit: AAAS Archives; photograph by John Howard Paine. Bottom:* "The savants of the American Association for the Advancement of Science greet their surprise new member," William Jennings Bryan. *Cartoon by Clifford Berryman,* Washington Star, *4 January 1925.*

LOOK WHO'S HERE!

F. R. Moulton, administrative secretary 1946–1948 (back row, left). *Credit: Reprinted with permission of Science Service, the weekly news magazine of science, copyright 1948.*

The Smithsonian Institution Castle, Home of the AAAS from 1907 to 1946. *Credit: Smithsonian Institution.*

Harry S. Truman addressing the 1948
Annual Meeting in Washington, D.C.
*Credit: Photography by UPI/Corbis
Bettmann, with permission.*

Arden House conference center, Columbia University. *Credit: Columbia University Office
of Public Affairs.*

Top: "Try it Again, Men, and be sure you get this answer." *Copyright © 1953 by Herblock in* The Washington Post. *Bottom:* 1515 Massachusetts Avenue, NW, first AAAS-owned headquarters building. *Credit: AAAS Archives.*

Project 2061, AAAS science education standards initiative. *Credit: AAAS Archives.*

Science for the People cartoon. *Copyright © by Tony Auth and the* Philadelphia Inquirer.

Warren Weaver, AAAS President 1954. *Credit: AAAS Archives.*

William Carey, AAAS Executive Officer 1975–1987. *Credit: AAAS Archives.*

Philip Abelson, Editor of *Science*
1962–1984. *Credit: The Carnegie
Institution of Washington.*

Dael Wolfle, AAAS Executive
Officer 1956–1970. *Credit:
AAAS Archives.*

Hesitation Over Traditional Priorities

Despite the apparent consensus over Arden House goals, the next two years saw little change in AAAS activities. In part, this happened because the AAAS tried to implement a major bureaucratic structure of committees and subcommittees to address the Arden House proposals, a structure that collapsed of its own weight before it ever became functional. In part, the delay was caused by personnel problems that occupied major blocks of time for the AAAS officers and board of directors (which replaced the executive committee of the council in 1952, when a new constitution attempted to clarify the difference between the policy-making function of the Council and the administrative function of the board). But in part, the delay was also caused by conflicts within the AAAS over the interpretation of the various Arden House proposals, especially the call for fewer specialized sessions at annual meetings.[27] After the 1948 "synthetic" meeting, most meetings had retained a substantial number of sessions arranged by specialized societies (acting in their capacities as AAAS affiliates). While some sessions had attempted to provide overviews or interdisciplinary cuts through science, many members felt these sessions usually consisted of reviews or mere restatements of existing knowledge, rather than the cutting-edge new knowledge for which they came to meetings.

At first, Administrative Secretary Howard Meyerhoff worked with the AAAS board to move forward on the Arden House proposals. But as time went by, Meyerhoff came to regard the Arden House proposals as direct criticisms of his own administration. At the association's annual meeting in St. Louis in December 1952, matters came to a head. The turnout was disappointingly low (fewer than two thousand), because of "the impression in many quarters that the Association already has decided to de-emphasize, if not to discontinue, its annual meetings." At the meeting, Weaver, newly selected as president-elect of the association, delivered a ringing address calling for more active attention to the Arden House proposals (the kind of address that drew letters praising Weaver for "the way in which you asserted immediate leadership to make the AAAS the spokesman, the champion and defender of science"). And, in a newspaper interview, Weaver and AAAS president Edward U. Condon strongly criticized the existing AAAS activities. A few weeks later, Condon and Meyerhoff would begin an exchange of angry memos about AAAS operations, establishing a particularly venomous personal relationship. Showing how national politics could influence organizational operation, Condon was angry in part because of comments Meyerhoff had made to a reporter regarding Condon's ongoing battles with the House Committee on Un-American Activities.[28]

Meyerhoff could contain himself no longer. In an editorial in *Science* in February 1953, he asked: "Where else but at a AAAS convention, can engineers, biologists, psychologists, industrialists, physical scientists, and public leaders assemble to consider Disaster Recovery? Or the Interface of Land and Sea? Or Problems of the Pacific Rim? It is not the Association that lags, but those who fail to comprehend the scope and impact of its current program. Intellectual bankruptcy and deterioration will indeed set in if the AAAS turns from programming important science merely to ballyhooing the importance of science."[29]

Several officers worked to bridge the gap between Meyerhoff, Weaver, and Condon, but the damage was too great. Meyerhoff and his assistant, *Science* executive editor Gladys Keener, resigned at the end of March 1953. The administrative turmoil that engulfed the AAAS when Meyerhoff left, as well as the general social turmoil caused by the Korean war, pushed all new initiatives off the list of immediate priorities. Until a new administrative secretary, Dael Wolfle, and a new editor for *Science* were hired at the end of the year, the association could not move forward. Wolfle had been secretary of the American Psychological Association (which had its offices in the AAAS buildings at the time), and was responsible for one of the first major reports on postwar training for scientists. As a result, he did not begin full-time work until mid-1954—shortly after which the new *Science* editor resigned, again upsetting workloads. Not until 1956 and the arrival of Graham DuShane as editor of *Science* did the AAAS settle into a normal routine again.[30] Thus the early 1950s were essentially lost years in AAAS history, reflected in several years of zero growth in the membership numbers.

Despite the AAAS's inability to pursue any new initiatives, its members continued to be concerned about the relationship between public policy and science—by which, in the heart of the McCarthy era, they often meant public appreciation of the value of free speech and inquiry. As academics, scientists were acutely aware of the chilling effects of Senator Joseph McCarthy's attacks on free speech; at least three AAAS officers had been called before the House Un-American Activities Committee. The same issue of *Science* that carried Meyerhoff's bitter editorial also carried Kirtley Mather's presidential address on the "Common Ground of Science and Politics," with a ringing defense of open scientific communication. According to the *New Republic*, Mather equaled Albert Einstein in the public's recognition as a scientist who spoke out on social issues.[31] Also that week, Ralph Rohweder, one of A. J. Carlson's colleagues in the letter that prompted the Arden House conference, wrote to Weaver worrying about the political persecution of scientists, which he described as "the problem of harassment of scientists in times of political anxiety." He proposed a national commission on "Intellectual Freedom and National Security," or, failing that, a National Science Week. He saw both as expressions of the philosophy that it was appropriate for science to claim special competence in the problems facing democracy.[32]

Nor was Rohweder alone. In December 1953, the American Physiological Society urged the AAAS and the National Research Council "to consider the possibility of undertaking a large scale program of public education concerning the importance and necessity of intellectual free enterprise for the development and utilization of science in the United States in the public interest." Though the AAAS board expressed interest, it declined to pursue the initiative.[33]

When Dael Wolfle arrived as AAAS's new administrative secretary (a position later recast as that of executive officer) in 1954, the AAAS board turned over all responsibility for Arden House to him. Wolfle found general agreement that the commitments to interdisciplinary science and synthetic programs were good ones. But when the board rejected Wolfle's plans for a new set of Arden House committees, he "turned 180 degrees." At his request, the board gave up the idea of creating a unified Arden House policy and instead opted to use the Arden House statement as a guide against which individual new initiatives might be judged. Wolfle also suggested—and the board apparently accepted—that the board act "non-democratically." That is, the members should do what they thought was right, and not wait for polls of what the association members might want.[34]

With the commitment to Arden House reemphasized and the implementation of policies entrusted to the existing management structure, the obstacles to AAAS change were successfully overcome. Arden House "was no longer treated as something in addition to or parallel with the rest of the Association's responsibilities, structure, and activities. Instead it became an integral part of AAAS planning and activity," Wolfle later wrote.[35]

Coordinating the Community

For the first one hundred years, the AAAS had achieved its goals most clearly through the annual meetings and the association with *Science*. Wolfle's integration of the Arden House statement into AAAS activities marked a dramatic shift. The history of AAAS since the mid-1950s has been one of new structures for the association's publications, periodic seminars for the mass media and other groups, popular book series, and changes in the structure, content, and operation of the annual meetings. In addition, members and especially staff have developed new initiatives in international science, scientific freedom and responsibility, science education, and other areas new to the association. While the AAAS grew in each of the periods covered by earlier chapters in this history, in its hundredth year, it would have been recognizable to its founders. By 1998, celebrating its 150th anniversary, the AAAS had become a fundamentally different institution. As time went by, changes in personnel, resources, and social conditions gradually reshaped the meaning that AAAS leaders drew out of the Arden House statement. What is striking about these changing meanings is that the tide of AAAS activities over the postwar years ebbed and flowed in response to funds, personnel, and internal politics, rather than in response to changing ideological statements.

Implementing New Policies

When Wolfle first arrived, he found that the AAAS's major asset, *Science*, was in deep financial trouble. A decade of rotating editors and leadership-by-committee had followed the death of James McKeen Cattell. The years had taken their toll. Circulation and advertising were stagnant. At the end of 1954, Wolfle heard that *Scientific American* was considering publishing its own weekly magazine of science and feared that it would sound the death knell for *Science*. To forestall competition with *Science* (which the AAAS considered the major weekly publication for science), Wolfle worked out a joint venture agreement under which *Scientific American* would have undertaken much of the work of producing *Science*—and much of the risk of marketing and publishing it.[36] But *Scientific American*'s board of directors, just then paying off the last of the debts incurred in that magazine's 1948 recasting from an inventors' magazine to one about scientific inquiry, insisted on changes to the agreement so unfavorable to AAAS that Wolfle's "first account of them was in a call from [*Scientific American* publisher Gerard] Piel to say that the deal was off; we could not possibly accept what his co-owners were offering." Nonetheless, Piel and his partners continued to work with Wolfle, and newly appointed editor Graham DuShane, to develop *Science* into a better and more successful magazine. In particular, the *Scientific American* staff helped Wolfle and DuShane recruit Earl Scherago to become advertising representative for *Science*, a relationship that would last for more than thirty-five years.[37] The new flow of funds provided financial stability to *Science*.

By 1956, with the problems of the flagship journal *Science* well in hand,

Wolfle and DuShane could begin to deal with the problems of the *Scientific Monthly*, which had been created in 1915 by Cattell as a vehicle to "review scientific progress and advocate scientific, educational, and social reforms." It was not intended to make science more popular or available to a broader audience—indeed, Cattell had sold his *Popular Science Monthly* to get away from that field. Like the new *Scientific American*, the *Scientific Monthly* was intended for an audience already informed about science. Its editor described it as "the *Atlantic Monthly* of science."[38]

But in the mid-1950s, the magazine was not financially healthy, and its audience was poorly defined. Unlike *Scientific American*, which clearly understood that its audience consisted of the industrial and technological elite whom advertisers wanted to reach, the *Scientific Monthly* had a much more nebulous audience of science teachers, retired scientists, and others on the periphery of science. The changing postwar economics of magazine publishing (in which advertising sales directed toward identifiable and coherent audiences provided more and more of a publication's revenue) worked against the *Scientific Monthly*, as did the growth of *Scientific American* and the health of the honorary scientific society Sigma Xi's *American Scientist*. None of the possible ways to revitalize the *Scientific Monthly* intellectually could overcome financial and organizational obstacles. In December 1956, the AAAS board voted to merge the *Scientific Monthly* into *Science*, with the hope that one issue a month of the new magazine would be devoted to the more general articles formerly carried by the *Scientific Monthly*. The merger became effective at the beginning of 1958, and circulation of *Science* immediately grew more than 50 percent, from 38,000 to 61,000. The plan to include philosophical and historical articles in *Science* failed, however. Though the board expressed its regret, it did little to follow up on the problem. Other activities devoted to public understanding of science removed the urgency of having a journal in any way devoted to that subject.[39]

Taking on Education

While the internal politics of the early 1950s had been dramatic, the AAAS itself had not been especially productive. The second half of the decade, however, was a much more energetic time. Wolfle's commitment to Arden House principles, combined with his pragmatic decision to implement those principles on a case-by-case basis, led to many new initiatives.

Some of the earliest new programs focused on science education—an area in which the AAAS had sporadically expressed interest, but in which it had failed during the Cattell years to create sustained programs. Since 1945, the AAAS had been home to the "Cooperative Committee" (known by that name, though the words "on the teaching of science and mathematics" were formally appended to the label), a group started by several other scientific organizations. This committee helped produce several important studies and reports on school curricula, teacher education, and related topics, including a summary requested by one of President Truman's chief advisors. Although the 1950 creation of the National Science Foundation, which took up education issues, removed some of the urgency for the Cooperative Committee, a 1954 reassessment still generated many ideas for improving science education.[40]

To implement these ideas, the AAAS overcame its long-standing ambivalence about accepting major funding from outside sources. Leaders like Wolfle had experience with funded research and were less concerned than earlier leaders had been about the potential loss of independence that such private foundation or government funding might bring. Weaver, who had played a major role in the creation of modern molecular biology by targeting donations from the Rockefeller Foundation, also understood the power of philanthropy as a tool for advancement.[41] By accepting outside funding, AAAS mirrored the changes in institutional finances sweeping through American science in its postwar expansions.

Thus in 1955, with a $300,000 grant from the Carnegie Corporation of New York, the AAAS hired John Mayor, a mathematician from the University of Wisconsin, to create its education department (initially called the Science Teaching Improvement Program). Over the next twenty years, Mayor would coordinate a series of projects designed to move science education away from "teaching" and toward "learning." He organized and sponsored conferences, including one that brought together the people who would later form the School Mathematics Study Group, progenitors of the "new math." He consulted regularly with universities around the country to improve their training programs and helped institute a variety of experimental systems to help teachers teach science.[42]

In 1956, the AAAS board held an extended discussion on "the problems of science education" and the complex implications of those problems for all of the association's activities.[43] In later years, many internal AAAS reports would cite those discussions as a watershed in how the association evaluated new programs. Then, in October 1957, the Soviet Union successfully launched the first artificial satellite, *Sputnik I.* After twelve years of Cold War, including the venom associated with McCarthyism and Red-baiting of the 1950s (ironically, Senator McCarthy himself died in 1957), Sputnik crystallized all the American fears about Soviet ability to use science and technology to its advantage. Within months, Congress had authorized the National Defense Education Act of 1958, which provided dramatically increased spending for science and technology education. The AAAS, with its existing programs and networks for improving public education about science, became one of the groups to which government agencies, congressional committees, and others turned for advice. The association's Washington location helped it function in this informal network, as its staff and leaders could easily attend meals or meetings with other national players. Many of these meetings were informal ones, or involved the AAAS in no role other than as facilitator or referral network. But other activities were more formal (and funded). Working with several other organizations and with state education directors, for example, Mayor helped create guidelines for teacher preparation in science, which in the early 1960s were widely adopted around the country.[44]

While many AAAS members contributed to the education projects, participation was not limited to the membership. The AAAS achieved its impact by serving as a catalyst for the general scientific community. Traces of the AAAS were sometimes hard to identify in the final outcome, but many of the developments in science education of the late 1950s and early 1960s might not have happened without the AAAS. Moreover, the education projects of the 1950s foreshadowed another trend that endures in today's AAAS—the increasingly "professional" nature of AAAS activities, with the use of grant funds from government agencies or philanthropic foundations to hire professional staff to carry out specific AAAS programs. While AAAS members serve as advisors, contributors, and sometimes even leaders for these projects, they depend for their day-to-day work on the headquarters staff.

For example, one of the most enduring initiatives in science education was the hiring in 1955 of Hilary Deason to create a "traveling library" of science books for high schools. In the first year, 12 sets of 160 books about science were sent to sixty-six high schools without their own science libraries. By 1963, when the program was transformed into a published *Guide to Science Reading*, nearly six thousand schools around the country had used the collections. The traveling libraries again showed the potential of AAAS to serve as a catalyst. When a traveling library for elementary schools was created, Deason insisted (based on experience with the high school libraries) that it go only to schools with a central library, to ensure that a professional librarian would be available to help manage the collection. "The AAAS hoped that this action would stimulate the acquisition of central libraries by those elementary schools that lacked such facilities," the *AAAS Bulletin* reported, and "as a matter of fact, some schools did establish central libraries in order to be admitted into the program."[45] The NSF stopped funding the program in the early 1960s because "widespread distribution of traveling libraries has so encouraged the acquisition of science books by science libraries that further [NSF] support of this activity is no longer necessary."[46]

In 1963, after producing a series of mimeographed annotated bibliographies, Deason published *A Guide to Science Reading*, focusing on widely available paperbacks. "Within 24 hours of the receipt of a review copy of [the guide], I was able to answer, quite specifically, two requests for information that might have left me blank-minded," noted the well-known science writer Isaac Asimov. "Coincidence, perhaps, but I take it as significant evidence in favor of the endless usefulness of the Guide as a reference to inexpensive reservoirs of knowledge, something of great value, surely, to every teacher and student of science." In 1964, the traveling libraries and annotated book lists became *Science Books* (later, in 1975, *Science Books and Films*), a quarterly magazine for librarians and others produced by professional staff at AAAS, with the aid of AAAS members across the country who serve as reviewers. *Science Books and Films* continues to serve as a major source of information for purchasers of scientific materials for educational purposes.[47] (In 1998, the AAAS began applying the SB&F model to Internet-based Web sites, in a Science Netlinks project funded by MCI).[48]

Science and Politics

From the beginning of the 1950s, AAAS leaders had become more and more concerned about the interactions between science as an institution and government policies. They recognized that this implied a change from earlier days in the association's life. Traditionally, AAAS presidents had used their annual addresses to sum up recent developments in their own fields. But in the postwar years, as government relations increasingly occupied the attention of scientific leaders, AAAS presidents used their platform to address broader issues. Giving his presidential address in St. Louis in 1952, Kirtley Mather noted that "instead of discussing some specific problem or some notable progress within the field of geology, my special department of the physical and biological sciences, I shall consider certain of the broad problems arising from the impact of science on modern life."[49]

In particular, for Mather (who had strong Marxist beliefs, though he was apparently not a member of the Communist Party), the crucial issue was the independence of science from politics. Government agencies depended on science, "but this is a two-way street. Scientists are increasingly dependent upon politicians. Approximately 35,000 specialists in the physical, biological, agricultural, and engineering sciences are now employed in government laboratories and research facilities, working under the supervision of politicians." He noted the growing use of loyalty oaths, censorship, and other aspects of what historians and political scientists now call "the security state." It would be "bad enough if the harmful policies and practices were applied only to the scientists working on projects that have security implications," Mather said. "Actually, they are extended far beyond that relatively small group. Political orthodoxy, rather than mere technical competence, has been accepted as a basic qualification in many academic institutions and industrial laboratories. . . . Administrators dare not risk charges that might be made by Congressional committees, or by radio and newspaper commentators, that they are employing 'red' scientists."

At a time when such prominent researchers as Linus Pauling were having their passports withheld, Mather defended the practical needs of scientists in high moral terms:

> Intellectual freedom involves the free interchange of information and ideas among scientists not only within our country but also between and among those of all nationalities. Among the most stimulating factors in scientific progress are the international gatherings of specialists in the various scientific disciplines and the visits of foreign scientists to the centers of research and development in the United States. Few persons, other than the scientists themselves, are aware of the tremendous indebtedness of American technology to scientific research prosecuted in other countries. Freedom to travel may not be one of the justly celebrated "Four Freedoms," but for the man of science it ranks at least as high as any other.[50]

Years later, many scientists would remember Mather's courage in making such statements at the height of McCarthyism. Mather's concerns about intellectual independence were seen as prescient just three months later, when political leaders in the Department of Commerce tried to force the director of the National Bureau of Standards to withdraw a report condemning as useless a battery additive marketed by a California company. The incident led to the forced resignation—and then reinstatement—of the NBS's director, Allen V. Astin.[51]

A more immediate issue that was beginning to absorb the nation during the 1950s also engaged the AAAS: racial segregation. In 1951, *Science* had published a letter from mathematicians at Fisk University, a black institution in Nashville, protesting their exclusion from a Mathematical Association of America dinner in a segregated hotel. Two years later, the Council for the Advancement of Negroes in Science urged the AAAS to use its affiliate structure to encourage desegregation throughout the scientific community; although Weaver responded with a letter, the association took no action.[52] Then, in 1955, matters came to a head when AAAS scheduled its December meeting for Atlanta. The association understood some of the problems of symbolism and practicality inherent in a meeting in the segregated South, but hoped to reach out to Southern scientists both black and white and to send a message by insisting that all sessions and social events be completely open to all races. Nonetheless, many members were outraged. Clifford Grobstein, a prominent biologist, resigned in anger over the explicit statement in hotel reservation blanks printed in *Science* showing that blacks would have to reside in different hotels than whites. A series of letters (many from anthropologists) protested the decision to meet in Atlanta, arguing that the difficulties faced by black scientists outweighed the potential benefits. Prominent anthropologist Margaret Mead, then newly elected to the AAAS board, served as an intermediary between the anthropologists and the board, ultimately siding with those who believed that the personal experience of segregation by Northern white scientists would solidify opposition to segregation. The meeting was held as scheduled and without incident (though it was boycotted by the anthropology section).

The belief that personal experience might shape people's opinions apparently bore fruit, according to historian Carleton Mabee, who wrote that "the distinguished physiologist Detlev Bronk, a recent president of both the AAAS and of Johns Hopkins University, stood in the rain in front of a downtown Atlanta hotel, furious because he could not get a taxi to take him to an AAAS session at the black Atlanta University. The white taxis could not take him to a black neighborhood and the black taxis could not pick up a white passenger. Bronk's realization of the utter folly of such a system . . . transformed him into a strong anti-segregationist." Bronk supported negotiations later in the meeting to have the AAAS Council vote for a resolution declaring that "American science does not place a regional educational mission above fundamental respect for the person and personality of human beings." The AAAS did not meet in the Deep South again until the 1990 New Orleans meeting.[53]

By 1958, the concerns about political issues reached a crescendo, and AAAS hosted a Parliament of Science to discuss issues of public policy. The organizers of the Parliament minced no words about their view of the importance of science; their report in *Science* began: "Power has always been sought avidly. Sometimes it has been used disastrously; often it has been used wisely. How the United States . . . shall ensure that knowledge (the age-old synonym for power) will be used for the benefit of mankind in general and its citizens in particular [is] among the most important questions before the American public today."[54]

A three-day meeting involving about one hundred scientists nominated by the AAAS Council and affiliated societies reviewed and made recommendations on a range of topics. On some issues, the conclusions were foregone and self-interested, such as a request for more support for science, especially basic science. But not always. The parliamentarians understood and rejected the narrowness of schemes like the then-current suggestion that the government establish a "West Point of Science." In yet other cases, the participants expressed ongoing concerns that seem as current today as they did then: "It is recommended that scientists themselves improve the presentation of papers by use of clear, vigorous English with correct technical terms and nomenclature, and improve the quality of what is communicated by supporting critical and responsible editorial policies in scientific periodicals. To this end we urge more systematic attention to expository writing on the part of graduate students in science."[55]

As with so many AAAS activities, this one yielded no specific outcomes, and apparently merged with the many other reports of concern about the relationship between science and society expressed in the months after *Sputnik*. But unlike in earlier years, when AAAS leaders were not necessarily active in AAAS affairs, the Parliament of Science represented a successful attempt by those scientists who were also leaders outside the association to use it as a forum in which to address issues of broad community concern. AAAS leaders during the postwar years were often deeply involved in association affairs both before and after their formal appointments as officers. Presidents during the late 1950s included Weaver (shortly before he began serving on the National Science Board, the governing body of the National Science Foundation), Paul Klopsteg (a physicist who spent nearly twenty years as a board member), and Chauncey Leake (a widely known pharmacologist and historian of science who would contribute articles to the program books for AAAS annual meetings for another decade after his presidency).

In the late 1950s and early 1960s, as the postwar concerns about the implications of scientific work spread from nuclear physicists to other researchers, the AAAS, with its pan-science orientation, became a comfortable, if not always welcoming, home. A few people, most notably Weaver, Mead, and the biologist Barry Commoner, became especially active in AAAS affairs and used the AAAS forum to raise issues of scientific responsibility. Especially for Commoner, who was an early opponent of nuclear weapons, the AAAS provided a platform unavailable elsewhere. Thus the AAAS led the way in making issues of science and human welfare common currency in the scientific world. The ability of the AAAS board to create committees addressing major issues provided a mechanism for these activities, a mechanism used throughout the postwar years. In 1957, an Interim Committee on the Social Aspects of Science, led by Commoner, reported that "there is an impending crisis in the relationships between science and American society. This crisis is being generated by a basic disparity. At a time when decisive economic, political, and social processes have become profoundly dependent on science, the discipline has failed to attain its appropriate place in the management of public affairs."[56]

Three years later, this committee had been transformed into the Committee on Science in the Promotion of Human Welfare, which noted in its initial report that "with each advance in our knowledge of nature, science adds to the already immense power that the social order exerts over human welfare. With each increment in power, the problem of directing its use toward beneficial ends becomes more complex, the consequences of failure more disastrous, and the time for decision more brief."[57]

Soon this committee turned its attention to specific issues. At the December 1961 Denver meeting, it issued a "statement of conscience" about the relationship of science and war: "War is today a social problem of catastrophic force and overshadowing urgency. The basis of war is power, and power is a product of science. Science is therefore deeply involved in this problem, and scientists have a particular duty toward its solution." Arguing that physics, chemistry, and biology had made possible a host of new weapons whose power could never be retroactively assessed because of the catastrophic devastation they would cause on first use, Commoner and his colleagues called "for the establishment of a new collaborative science, *the science of human survival*, which will apply the full strength and wisdom of all the sciences to the solution of the crisis created by the obsolescence of war." For these scientists, the AAAS was not simply a forum, but a prominent pulpit from which to promote their goals. "The present danger of war derives from the powers of science and decisions of society," they wrote. "Its resolution also depends on those agencies. It lies within the power of science to illuminate the self-destructive nature of modern war and to discover new social inventions to replace it."[58]

Observers greeted the report with excitement. "The opinion of a AAAS committee, to be sure, is not an 'official' AAAS position," wrote *Washington Post* reporter Victor Cohn. "But . . . it seemed to mesh with general AAAS thinking. For the first time, it seemed, a large general organization of American scientists may be displaying both aroused conscience and aroused determination to move past war as a social tool."[59]

Continuing to Meet and Publish

The presence of Cohn and other science writers at the annual meetings was by now routine, and the National Association of Science Writers had become a regular supporter of AAAS initiatives in public understanding of science. At the December 1961 meeting in Denver, the young astronomer Carl Sagan was part of a panel on extraterrestrial biochemistry. Walter Sullivan of the *New York Times* asked, "Has anyone ever suggested that dead seeds of life from other planets might become templates for life on this one?" Sagan, "who enchanted reporters with his wit and rapid-fire responses," shot back at Sullivan, "So far as I know, no one has; perhaps you might like to write a paper on it."[60] In the years to come, reporters who covered AAAS meetings would be identified as an "inner club," whose commitment to scientific ideals often mirrored that of the scientists they covered, and whose decisions shaped science reporting throughout the year.[61]

The meetings themselves continued the eclectic mix of specialty-science and broad thematic sessions that become the compromise in the years after the Arden House meeting. But by the early 1960s, the trend was clearly away from specialized sessions, and AAAS leaders had accepted that trend as inevitable. In 1963, Alan Waterman (recently retired as founding director of the National Science Foundation) noted in his president's letter that most of the large specialty societies, including the American Chemical Society and the American Physical Society, no longer met with the AAAS. In addition, even specialty sessions organized by individual sections "create problems both of management and of attendance," he said. As a result, AAAS meetings now encouraged "more and more symposia on broad subjects of national interest. . . . These have been highly successful and I believe the AAAS is generally acknowledged to constitute an excellent forum for such ventures."[62]

Since the 1930s, AAAS had published the proceedings of many symposia (both technical and general), and these volumes were influential in carrying discussions begun at AAAS meetings to wider audiences. In areas such as industrial science, the publications helped indicate the association's interest in expanding its vision of what to include in "science." Unfortunately, reflecting the shift in science away from books and toward journals, many of the volumes failed to sell well. In the mid-1970s, during a period of financial stringency at AAAS, the original symposia series died. Over the next two decades, the association experimented with various forms of self-publishing and arrangements with commercial publishers to ensure that the proceedings of important symposia at the annual meetings remain available for future reference. In addition, the association arranged for selections from special issues of *Science* on energy and other topics to be published in book form.[63]

The interdisciplinary and issue-oriented symposia were also taken up by the AAAS's regional divisions. As with the Pacific and Southwestern (later, Southwestern and Rocky Mountain) Divisions created during the first half of the century, the continuing problems of travel and communication across the breadth of the American continent spurred the creating of the Alaska Division in 1951—many people recognized that scientists from that region could not easily travel to the annual meetings held in the lower forty-eight states. Moreover, the particular scientific issues of certain regions lent themselves to the kind of interdisciplinary activity that AAAS leaders wanted to foster. The most dramatic example had occurred in 1954, when the Southwestern Division proposed an international meeting on arid lands. Just a year later, after frantic organizing and international fund-raising, the first International Arid Lands meeting was held in Albuquerque, New Mexico, with UNESCO sponsorship and representatives of nearly thirty countries. The stimulus affected both the field and AAAS; for a generation afterward, the AAAS as a whole sponsored a committee on arid lands that served as a major locus for organizing conferences and coordinating research efforts sponsored by various public, private, and United Nations groups. Similarly, the Alaska Division proved to be a successful forum for bringing together scientists from the United States and Canada to address common concerns, leading to its 1982 renaming as the Arctic Division. While the initial concern about the costs and difficulties of transcontinental travel eased as the century wore on, the ability to focus on regional issues proved productive, and in 1984, the Caribbean Division was formed to provide yet another geographic focus. In addition to regional focus, the annual meetings, workshops, and field trips now sponsored by the regional divisions often offer opportunities for graduate students and other new researchers to begin their participation in professional activities.[64]

As the AAAS moved away from meeting programs directed by the disciplinary societies that were its affiliates, its ability to look at issues that crossed disciplinary boundaries was proving to be its greatest strength. To leaders like Waterman, this change had even broader implications for AAAS priorities, showing the enduring strength of the Arden House reforms. "It is my personal view, and that of many with whom I have consulted," Waterman wrote, "that a major and unique function of the AAAS is unquestionably to increase pubic understanding and appreciation of science in general and of current scientific issues in particular. . . . There has never been a time when great issues of national and international policy have been so closely linked to the progress of science and technology. In fact, that very progress in certain fields has itself become a subject of public policy. How great is the need, therefore, for the public to have some notion of the significant trends and developments in science and technology and of the issues that bear upon them."[65]

Long-range planning and articulation of "public understanding of science" goals had been part of the Arden House outcome. But as in other areas of AAAS interest, and despite a series of memos, grandiose plans, and grant applications, few new initiatives were created by the staff in the 1950s. At Weaver's urging, the board appointed a Committee on Public Understanding of Science in 1958 to make public understanding a permanent part of AAAS's culture, but the new committee still could do no more than provide moral support for ongoing activities.[66] By the early 1960s, the need for a permanent staff to create and manage meaningful AAAS initiatives had become clear. Mayor had proved the value of staff in education. Now the AAAS needed a single person to coordinate public understanding of science activities. In March 1961, it hired E. G. Sherburne, Jr., who had worked extensively in educational television, the area in which AAAS leaders thought their programs might have the most impact.[67] Over the next few years, Sherburne brought in funding from the Sloan Foundation (where Warren Weaver was now executive vice president) to help establish opportunities for discussing science on television, to support research on public understanding by Stanford University communication professor Wilbur Schramm and his students, and to enlarge the operations of the AAAS's annual press room. Cooperation with television producers was so successful that Dael Wolfle seriously contemplated opening AAAS offices in Hollywood and New York to serve the television community. Sherburne also established a newsletter to reach out to the broad community that he thought should be involved in public understanding activities; by 1964, *Understanding* had a circulation of 3,300 people in journalism, television production, public information offices, industry, government, adult education, museums, and science.[68]

By 1964, reflecting a shift in public attitudes toward science, Sherburne began to suggest that an "understanding" of science was not necessarily the same as "appreciation" for science. Understanding did not mean simply a knowledge of certain facts, he said, but also of science as an activity and of the impact of science on society.[69] Sherburne was probably influenced by the stirrings of doubt about science that emerged in the early 1960s. Rachel Carson's *Silent Spring*, with its indictment of science's role in the damage done to natural resources, had been published in 1962. Scientific groups opposed to the testing of nuclear weapons had coalesced into the new Scientists' Institute for Public Information in 1963; in its earliest incarnations, SIPI (led by Barry Commoner) was a group distinctly dissatisfied with the vision of science as necessarily the provider of answers for social problems. The years 1962 and 1963 marked the end of the unbridled postwar enthusiasm for science. The early environmental and antinuclear movements began to grow as political and social concerns developed suggesting that science, rather than solving problems, might be creating them. Science came under attack. The moral certainty in science that had motivated the immediate postwar leaders began to waver.[70]

But not universally. To some AAAS leaders, even a tame shift away from a central focus on and belief in the absolute value of the scientific perspective was unacceptable. Polykarp Kusch, the chairman of the association's Committee on Public Understanding of Science, angrily resigned in 1964, complaining that Sherburne was "an assertive Madison Avenue type," who was hard to work with and who was taking the public-understanding program in the wrong direction.[71] For Kusch and others, popularization was not simply an honorable duty; popular science had to serve as the weapon for fending off attacks. Scientists began to circle their wagons. Two incidents at the AAAS suggest the tensions that these new attacks on science imposed on the association.[72]

During the late 1950s and early 1960s, *Science* had finally replaced Cattell's old "science news" gossip section with a lively "news and comment" section. Coupled with editorials from a variety of authors, the magazine had become a forum for issues of concern to the scientific community. As such, it had begun to attract readers from outside the scientific community. Suddenly, it appeared, nonscientists would be privy to disagreements within the scientific community; some scientists feared that this would undercut the image of authority and certain knowledge that scientists used as part of their defense against attacks. In a 1963 letter to *Science* editor Philip Abelson (who had taken over from Graham DuShane the year before), National Academy of Sciences president Frederick Seitz worried that

Science is attempting to establish policy by means of an extensive open public debate through editorials and letters in which the scientific community is invited to express whatever views, extreme or otherwise, individuals may have. . . . One wonders if [AAAS] would not render a greater service to the country by creating a body . . . which would attempt to review the opinions of individual scientists in a constructive way and then use the pages of *Science* to summarize the results of these deliberations after they have been digested. A great deal is at stake here for the entire scientific community, and hence for the nation. One hopes the results of the work of the AAAS will really work for the advancement of science.[73]

Seitz's challenge, coming from the most influential and respected body in the scientific community—and AAAS's longtime rival for leadership in the community—could not be ignored. But the tension between the "elite" National Academy of Sciences and the "more democratic" AAAS was clear in the AAAS leaders' response: in a democracy, debate must be open. "It is our faith," said a statement prepared by Wolfle for the board of directors, "that both science and democracy are most likely to flourish under conditions that permit free discussion and clear presentation of opposing points of view. *Science*, therefore, does not pretend to be the authoritarian voice of scientists generally, of the members of the AAAS, or of the Association's officers. It is a medium for the presentation of scientific findings and theories, the analysis of scientific problems, and the discussion of issues that are of vital concern to the advancement of science."[74]

This commitment to the identity of science and democracy was a common theme in the postwar years as it had been throughout AAAS history. But the second incident at AAAS revealed the limits to which this identity could be pushed. Also in 1963, *Science* news reporter Daniel Greenberg (a former *Washington Post* reporter) suggested to the AAAS that it create a magazine to cover the general topic of science policy. Although Greenberg and others at the AAAS worried about whether a new magazine would have sufficient commercial appeal and intellectual focus to survive, Greenberg succeeded in getting endorsements for his project from the AAAS Committee on Public Understanding of Science, and from leading AAAS members such as Margaret Mead and Barry Commoner.[75]

He also, not surprisingly, got support from Warren Weaver. Weaver thought that the audience for the new journal might well come from the scientific community, though it should be directed to "the general public . . .

[which] includes scientists, of course." Weaver anticipated a start-up time of more than five years before the magazine would be successful, but this did not affect his enthusiasm. The role of the new journal, he said, "is not 'the interpretation of science to society.' It is the identification and illuminating of the issues of interrelationship between science and society." In what turned out to be a particularly prophetic warning, Weaver repeated his concern of more than a decade earlier that "it would be fatal if it started out with the arrogant premise that scientists are always right (on public as well as scientific matters)."[76]

Greenberg put together a prototype issue, using articles he solicited and published also in *Science*. He thought the prototype had been well received, but the AAAS board expressed concern about the project's viability. And, though the board had just months earlier defended the appearance of debate in *Science*, it perceived greater dangers in deliberately spreading that debate outside the scientific community. Reflecting the growing concern about the image of science, several board members objected to Greenberg's highlighting of the phrase in one article that "science is too important to be left solely to the scientists." In an angry discussion sparked by the phrase, the AAAS board voted to terminate plans for the magazine.[77]

Together, the Kusch, Seitz, and Greenberg affairs help define how the interaction of politics, public understanding, and science's own institutional needs shaped AAAS in the 1960s. The AAAS was comfortable providing information about the substantive content of science, information intended to provide nonscientists with an appreciation of the benefits that society received from science. For AAAS members, "understanding" meant an understanding of the improvement of human life as a result of specific research areas in the basic sciences. Scientists continued to be motivated by their moral certainty in science, by their confidence that it provided not only *an* answer, but *the* answer to many of society's problems.

Serving Science

The programs and opportunities in science education and in public understanding of science represented new directions for the AAAS in the postwar years. In its earliest years the AAAS had served as a forum both for scientific research and for promoting the importance of science in America's emerging culture. But since the 1870s, the AAAS had devoted most of its energies to supporting and publishing scientific research. Now the focus on education, popularization, and policy meant a return to defining the "advancement" of science in much broader terms, especially including coordination with other scientific groups, as well as diffusion to and interaction with non-scientists. Moreover, at the institutional level, the new projects represented a shift in AAAS history from an organization of individual leaders to one of ongoing programs supported by full-time staff. This shift emerged as the number of scientists grew dramatically in postwar America. From about 250,000 in the early 1950s, the number of scientists and engineers employed in research and development had grown to more than 500,000 in the late 1960s, and nearly a million in the early 1990s (the latest period for which figures are available).[78] No longer could individual researchers, in their spare time, organize meetings or projects just drawing on their networks of friends and contacts. Instead, central offices with their ability to raise federal or private funds, provide stable office staffs, maintain mailing lists, and so on, took on a more prominent role in the organization of modern science.

In keeping with the shift from member- to staff-driven programs, perhaps the most important leadership came from the professional staff. This was especially true for psychologist Dael Wolfle, who was executive officer from 1954 to 1970. Wolfle oversaw the shift in AAAS's mind-set from inward to outward focus. He oversaw the growth of AAAS from a small organization operating out of rooms in the Smithsonian to one owning its own state-of-the-art office building in downtown Washington. In 1966, commenting on AAAS structure, AAAS president Alfred Romer noted that Wolfle was "a man of unusual ability—sage, level-headed, politic, and discreet." The AAAS staff agreed. On Wolfle's retirement, a public letter from John Mayor talked of the "very great respect and esteem" the staff had for Wolfle. "Your judgment and competence have never been questioned, not even on rare occasions when maybe they should have been. Your kindness, the generous spirit with which you have always given of your time, and your sincere interest in the welfare of the staff have been very important for us."[79]

One of Wolfle's most important achievements was to clarify the relationship between the AAAS and *Science*. In the decade after the AAAS had taken over the journal from the Cattell family in 1944, the association's administrative staff had also served in editorial roles for the journal. While this had served to integrate the journal more fully into the association's affairs, it had also proven to be unwieldy and distracting and contributed in some part to the tensions of the early 1950s. When biologist Graham DuShane took over as editor in 1956, Wolfle established a different model. As executive officer of the AAAS, he would serve as publisher of the association's journals. He was thus ultimately responsible for their success, their marketing, and (at the practical, everyday level) the maintenance of the subscriber list that was simultaneously the association's membership list. Under his leadership, for example, *Science* was redesigned to make it more attractive to advertisers. But for the most part, he willingly delegated to the editor the direction of the journal. The 1963 dispute with the National Academy of Sciences, by generating the statement endorsing *Science*'s editorial independence from AAAS, further ensured that the journal would have the credibility needed to climb to the top ranks of the scientific publishing world.

In the late 1950s, DuShane began introducing more news and policy material to *Science*, and when the *Scientific Monthly* was shut down, the association found that its more focused editorial efforts helped it attract more interesting and timely work. DuShane also introduced modern peer review to *Science*, soliciting manuscript reviews from a broad range of scientists, cutting reliance on the small though dedicated editorial board. Still, during the 1950s, *Science* suffered by comparison with *Nature*, where many of the earliest publications in the "new" molecular biology and biochemistry were appearing.

The importance of having a high-quality journal grew as the AAAS grew and the possibility of attending its meetings (whether national or regional) grew slimmer for many of its members. More and more, people considered the journal to be the primary benefit of their membership in the AAAS. In this context, another key individual, Philip Abelson, became editor of *Science* in 1962 when DuShane resigned to become dean of graduate sciences at Vanderbilt University. Abelson was already a fabled scientist, a student of E. O. Lawrence, who had done pioneering research before World War II, including discovery of the transuranium elements neptunium and plutonium. His own research interests covered chemistry, microbiology, geophysics, and nuclear chemistry.[80] By the early 1960s, he was director of the Carnegie Institute of Washington's Geophysical Laboratory and editor of the *Journal of Geophysical Research*. He maintained his Carnegie affiliation throughout his years at *Science*, moving up to the presidency in 1971. And even after stepping down as editor in 1984, he remained as deputy editor for engineering and applied sciences, a position he continues to hold at this writing. Abelson's strength was his intellectual vitality. "He rarely delved into a field, no matter how remote from his own background, without encountering new results or techniques that he found fascinating," wrote one of his colleagues. "He reserved particular enthusiasm for experimental results derived from pushing a technique to new limits and was quick to see the implications for further research or application. . . . To work beside Abelson was to watch how rapidly he grasped the essentials of a new scientific idea, turned it this way and that to see its implications, and probed it for its weaknesses."[81] Abelson was part of the elite group of physicists who rose rapidly in the World War II research effort, flowing easily into the new postwar councils of science and government.[82]

Abelson quickly made his mark on *Science*, adding to its role as a scientific journal by enhancing its ability to explore the interactions of science and other social institutions. He strengthened the news staff (a process begun by DuShane), making *Science* an important vehicle not only for learning about current research, but also for learning about government and industrial actions related to science. Under his leadership, *Science* made issues of pork-barrel funding, military technology, and other government funding of science major topics in the scientific community. (The growing role of the AAAS as a player itself among the institutions of science did lead to some conflicts. After a series of disputes about how to cover AAAS activities in a journal that was both published by the AAAS and editorially independent of it, "news and comment" reporter Daniel Greenberg left *Science* in 1970 to create his own widely read newsletter, *Science and Government Report*.) In later years, Abelson added a "research news" staff that provided readers with overviews of emerging and important scientific fields. Beginning in the 1970s, the AAAS published in book form collections of articles originally produced by the news staff on issues such as energy, heart disease, and health care regulation.[83] In a report to the AAAS board near the end of his first year, Abelson noted that "in the past five years, *Science* has retained the contents of a journal but has acquired some of the attributes of a magazine: a large circulation, a strong financial base, efficient printing arrangements, a writing staff, and an experienced editorial staff."[84]

Abelson also made his mark on the traditional, peer-reviewed side of *Science*. Under DuShane's leadership, the journal had emerged from the doldrums of the late Cattell years and the turmoil of constantly changing editors to become one of the major scientific journals in the world, though not yet a leading one. The increased prominence led to increased workloads, and delays between the acceptance of an article and its publication were becoming more common. Abelson continued DuShane's move toward a wide corps of manuscript reviewers, and introduced the telephone into the reviewer-recruiting process. In just a year, he reduced the median delay in research reports from four months to two months. He also struggled to increase the coverage of physical sciences (fighting the traditional emphasis in *Science* on biological articles), largely by using his network of contacts to solicit new articles. Abelson was also directing his staff to be more aggressive in editing and rewriting. "Because it has readers in many fields," he wrote, "*Science* must be oriented toward readers rather than authors."[85] In the 1960s, *Science* published key articles in emerging fields like continental drift and ecology. Continuing the AAAS's tradition of attention to conservation and environmental issues, in 1968 *Science* published Garrett Hardin's "Tragedy of the Commons," which originally had been delivered as the presidential lecture at the Pacific Division's annual meeting.[86]

The *Science* staff prided itself on addressing the challenges of rapid publication, which Abelson noted in his periodic reports to the AAAS board and membership. The system was tested in January 1970, when *Science* prepared to publish papers from the NASA *Apollo 11* Lunar Science Conference within a month after the meeting. Contributors were required to provide their manuscripts to the editors when they arrived at the meeting on 4 January. After initial screening, the papers were given to referees (most of whom were at the meeting), and reviews were received by 6 January, with comments and discussions with authors continuing through the end of the meeting on 8 January. Marked-up manuscripts were sent directly to the printer, and the 30 January issue containing the papers appeared on time.[87] Subsequent special issues appeared regularly, such as the scientific results of *Spacelab* in 1984, and papers on Comet Shoemaker-Levy in 1995.

By the late 1970s, *Science* was fully established as one of the two top general-science journals worldwide. Under Abelson's leadership, its circulation had more than doubled. It ranked seventh among 3,500 science journals for total citations, and scored in the top 2 percent of those journals for "impact," according to studies conducted by the Institute for Scientific Information.[88] When the popular science magazine *Science Digest* altered its cover logo to resemble *Science*'s, the association sued; the court decision in AAAS's favor ruled that *Science Digest* "differs from *Science* as a Philip Roth novel differs from a Shakespeare play, as Bo Derek does from Katharine Hepburn."[89]

As *Science* and the AAAS grew, so did the financial complexity of the organization. For most of its history, the AAAS's finances had been handled by its scientific leaders. It did have financial professionals on staff (particularly, during the postwar years, its business manager Hans Nussbaum, who began working at AAAS in 1945 and continued as a consultant into the late 1980s, well past his 1975 retirement). The board had sought out the advice of professional money managers, creating an investment committee. But none of those advisors had a sustained interest in the intellectual core of the AAAS. That changed in 1969, with the appointment of William Golden as treasurer.

Golden, an amateur radio enthusiast since 1922, had begun his career in the 1930s as an investment manager on Wall Street. After navy service during World War II, during which he invented gunfire devices, he was one of the original staff members of the Atomic Energy Commission. Later a key architect of the first office for science advice to the president in 1950, Golden was as well connected to the leadership of science in the postwar years as Abelson. He returned to investment management and corporate direction in the early 1950s, but remained extremely active as a government consultant and advisor. He cared deeply about the importance of science in public affairs, and shared the concern that the AAAS should reach out to the public. He remains treasurer and a major benefactor of the AAAS at this writing, and has also served as a leader of the New York Academy of Sciences and the American Museum of Natural History.[90]

In the 1960s, the AAAS was a leader in using television to advance scientific communication. Working with other societies, NSF and industrial funding, and a few educational TV stations (the precursors to today's public television network), in 1962–1963 the AAAS experimented with a series of shows about contemporary science. The following year, with additional funding, the series went national to more than seventy-five stations. Later in the decade, the AAAS arranged for major television networks to broadcast roundups of recent research from the annual meetings.

Much of the work on science education also came to fruition in the late 1960s with the production of Science—A Process Approach (S-APA). One of the first of the comprehensive "hands-on" curricula for elementary schools, S-APA was produced after nearly a decade of research and development by the AAAS's Commission on Science Education. Funded by the National Science Foundation and the Xerox Corporation, by the early 1970s it was being used by nearly 100,000 teachers and about three million students nationwide. The set of hands-on materials included about 1,500 distinct items (no one knew exactly how many), including balances, inclined planes, marbles, seeds, and animals. The curriculum stressed the process of science, including observing, classifying, measuring, predicting, inferring, interpreting data, and communicating.[91]

At one level, the new curriculum (and several developed by other groups at the same time) was successful. Controlled studies showed that students using hands-on, process-oriented approaches to learning science did improve their performance in a range of areas, from achievement and attitude through creativity. But many of the improvements were small, others were not statistically significant, and the curricula required substantial new costs and training for many teachers. By 1979, S-APA and similar curricula were being used by only about 10 percent of teachers. Nonetheless, the curricula were influential and contributed to the development of newer materials and approaches in the 1980s and 1990s, including the AAAS's Project 2061 and the *Benchmarks for Science Literacy*.[92]

Advancing Science in a New Context

As the 1960s progressed, the general social upheaval affecting all American institutions gradually made its presence felt within the AAAS. The end of the "Golden Years" of science (roughly the generation after World War II) were clearly evident in changing perspectives of AAAS presidents. In a 1962 "President's Letter" in the *AAAS Bulletin*, Paul Gross noted that "it has been largely the advance of science which has resulted in our ever more complex technological civilization with its new problems, its new and awesome weaponry. At the same time, there has been a phenomenal growth of our knowledge of nature through intensive scientific research that holds great promise for the future health and well-being of the human race, if wisely employed."[93]

Just six years later, after the assassinations of John F. Kennedy, Malcolm X, Martin Luther King, Jr., and Robert Kennedy, after the escalation of the Vietnam War, after the "Summer of Love" and the Watts riots, 1968 president Walter Orr Roberts was less sanguine: "Science and technology . . . entwine themselves ever more rapidly and ever deeper in the fabric of society. They affect the lives of us all in ways that increasingly dominate our choices. Our day in history desperately needs the optimism that is the tradition of science. Instead of optimism, our nation, blessed with unprecedented wealth, suffers anguishing doubts about whether we have wisely used our wealth at home and abroad."[94]

The following year, 1969 AAAS president Bentley Glass was even more pessimistic about the prospects for science, suggesting that information overload, economic constraints, technical specialization, and other factors would inevitably slow the growth of science. Moreover, reflecting the increasing concern about the effects of science, he argued that a "highly critical limiting factor in the growth of science is the lack of sufficient study of the long-term, unanticipated side effects arising from the introduction of insufficiently tested technological developments." Glass argued that scientists must work actively "to forecast as early as possible the effect of technological changes on society."[95]

Dealing with the End of Optimism

For the AAAS, the new, more skeptical attitude of the 1960s took some years to work out. As noted earlier, the 1963 attempt to create a magazine on science and public policy was quickly halted, yet the intertwining of science and politics continued to grow. In 1965, concerned about "the consequences of decreased federal support for training and research in the fields of science and health," the AAAS Council asked for a report on the problem. A committee led by Harvard political scientist Don K. Price found the concerns unfounded—they were based on rumors that science budgets would be cut to pay for the increased military budget needed for the Vietnam War. But as that link makes clear, federal support for science had now become so important that scientists could not ignore the political context in which even noncontroversial basic research operated. Moreover, while the rumors were not specifically true, the financial largesse for science did in fact begin to wane in the late 1960s.[96] In constant dollars, support for all research and development nearly quadrupled between the early 1950s and 1965. It then stayed essentially flat for the next ten years. Moreover, during those ten years, federal support dropped nearly 35 percent, meaning that the funding for R&D was increasingly tied to industrial needs. Even when federal funding rose in the late 1970s and 1980s, much of the rise came in defense-related funding.[97]

Two years later, when Price became the first political scientist president of the AAAS, he noted that many members feared his election meant that "AAAS was getting into politics." On one hand, Price acknowledged that "the AAAS—along with every other scientific association and every university in the country—is now in politics, in the sense that it must be interested in the role of science and education in public affairs." But, he said, that very interest "requires it all the more carefully to stay out of politics in the popular sense of the term—competing for political power." For Price, the proper approach was for the AAAS to continue to develop activities that addressed contemporary issues that would be of importance both for science and for politics. He noted, for example, that Connecticut congressman Emilio Daddario (who would later become an active leader of the AAAS) had proposed a Technology Assessment Board, "an effort to deal with the same problems faced by the AAAS Committee on Science in the Promotion of Human Welfare," and that Daddario's work on environmental pollution paralleled the AAAS Air Conservation Commission. Price also encouraged new initiatives such as seminars for congressmen and congressional staffers, which had recently been initiated under the "public understanding of science" program.

Many of these outreach activities lapsed in the late 1960s, after

Sherburne left to become the head of Science Service (publisher of *Science News* and coordinator of the national science fair competitions). But the interweaving of science and politics would not go away. The tumult of national culture and purpose that had been engaging the nation finally reached the AAAS during the 1969 Boston meeting in the week following Christmas. The meeting was the largest to date, with nearly eight thousand registrants. While the program organizers were proud that they had included topics like "Technology Assessment and Human Possibilities" and "Is there an Optimum Level of Population?" as major themes for the meeting, they were unprepared for the passion, venom, and vitriol that the meeting unleashed.

Student radicals had disrupted several recent scholarly meetings, including those of the Modern Language Association, the American Philosophical Association, and the American Historical Association. So AAAS organizers were not entirely surprised when a meeting held in Boston, "a city with the highest concentrations of radicals in the country, next to San Francisco," brought out protesters. They had even invited students to organize a series of panels on "The Sorry State of Science—A Student Critique," and had given the students (who called themselves the "Student Action Group") an office in the meeting headquarters. And, in the spirit of "stimulating a wider communication with the public concerning science, technology, and social change," the AAAS had waived its usual fees, opening the meeting to the public free of charge.[98]

But the Student Action Group was joined by the more radical Columbia University Ecology Action Collective. As a Boston reporter wrote in *Science*, "while ten Harvard and M.I.T. students carried placards silently around the Constitution Room of the Sheraton Boston (in front of WGBH-TV cameras), . . . the Columbia students were disrupting the meeting by shouting taunts at the panelists from the audience." In another session, a student screamed at panelists, calling them "war criminals" and asserting that true disarmament could come only through revolution. One former Columbia student grabbed the microphone from Charles Stark Draper of M.I.T.'s Instrumentation Laboratory and made a five-minute obscenity-punctuated speech.[99]

The spirit of rebellion affected more than just student radicals. "When a large crowd . . . found itself barred from a session on nutrition with Margaret Mead because the room was full, a matronly woman began shrieking at distraught convention officials," reported the *New York Times*'s Robert Reinhold, "wanting to know why the session could not be transferred to the nearly empty grand ballroom next door. 'I never thought I'd be an agitator,' she told a friend as her face turned purple with rage." In Reinhold's view, the woman "for a fleeting moment in the corridor of the Sheraton Boston Hotel . . . got a taste of the frustration that many young people and young scientists feel when dealing with what they think are equally exclusionary methods by which science policy decisions are made."[100]

Most important, the spirit of concern had also reached the elders of the association. As the *Washington Post*'s Victor Cohn reported in the *AAAS Bulletin*, "the angry kids were a prominent part of this AAAS meeting. . . . They contributed to an atmosphere in which the AAAS Council overwhelmingly (114–51) took a strong moral stand against continued use of possible mutagens 2,4-D and 2,4,5-T in Vietnam."[101] Although many observers considered the AAAS to symbolize the conservative, "establishment" place of science, the responsiveness of the organization to the forces for change signified its attention to emerging issues. The association had in recent years started committees on minority affairs and on the needs of young scientists. The resolution on pesticides in Vietnam was itself the direct outgrowth of an initiative begun in 1966 at the Pacific Division; that initiative had led to the AAAS Herbicide Assessment Commission, led by Matthew Meselson, a molecular biologist at Harvard, which in 1970 presented a report that one observer called "the greatest service the AAAS has ever performed for the human race."[102]

But clearly the 1969 meeting was a turning point for the association. The AAAS's first woman president-elect, mathematician Mina Rees, was selected by the council at the meeting (although Dael Wolfle believed that Margaret Mead, who had declined nomination several times, could have been elected many years earlier).[103] Within six months, Wolfle had stepped down (as previously planned) as executive officer after sixteen years of service. William Bevan, a psychologist whose work focused on sensory psychology and perception, took over from Wolfle, just as his own publications "began to emphasize the role that science and psychology played or ought to play in promoting human welfare."[104]

Bevan recognized the complexity of society's relationship with science. In a National Research Council address six months after assuming his duties, Bevan argued that the apparent disaffection of the public from the scientific enterprise was not entirely unjustified. "Our emphasis on excellence in individual performance has fostered a psychology of elitism that has made both our enterprise and our body of knowledge esoteric and increasingly inaccessible to the layman at all levels of society," he noted. He argued that "we have persisted in the view that science is value-free. . . . But one need only observe the keen competition that exists among one's colleagues in the pursuit of discovery, or hear their anguished cries as the shifting of research funding follows the waxing and waning of particular fields, to hold suspect the widely proclaimed neutrality of science." Moreover, he said, "the preoccupation with the primacy of creativity in individual investigations has prompted a circumscribed perception of the implications of one's scientific work."[105] Bevan apparently believed that it would be immoral for scientists to claim the sort of moral high ground so common in the years immediately after World War II.

For Bevan, the solution to the disaffection with science was "a need to increase the public's understanding of science." Unlike his Arden House predecessors, however, Bevan did not automatically add "appreciation" to the standard phrase. Instead, he argued that "first, there is the need for an assessment of our own attitudes. We must recognize the interdependence of science and society." Under Bevan's leadership, new blood reinvigorated the AAAS's public understanding activities. The new chairman of the Committee on Public Understanding of Science was Gerard Piel, publisher of *Scientific American*, one of the most successful new ventures in public communication of science in the postwar years. With the collaboration of other new COPUS members such as Harvard physicist and historian of science Gerald Holton, Piel had COPUS begin a series of studies investigating four "channels"—AAAS cooperation with teachers, TV-radio programming, science publications for students, and science kits.[106] More new ideas came in 1971, when Bevan hired James C. Butler, a former public information officer for Johns Hopkins University, to become director of communication programs at the AAAS. Butler's vision and responsibilities were broad, encompassing everything from AAAS symposium publications to wire services, magazines, and fine-arts FM radio stations.[107] Under Butler's direction, the association received nearly a million dollars from NSF for a wide range of public outreach activities, including new publications, research, youth activities, newsletters, seminars, travel grants, and a "National Center for Public Understanding of Science."[108]

As early as 1972, AAAS staff worked with producers at Boston's public television station, WGBH, to develop a new weekly science television program. Those plans came to fruition in 1974, eventually becoming the widely acclaimed program *Nova*. AAAS members read about the development of early shows before they were aired—one *AAAS Bulletin* featured pictures of a cameraman dangling from an airplane above the Colorado River—and the "AAAS News" section of *Science* featured the weekly schedule for *Nova*.[109]

Explosion of New Initiatives

Bevan's tenure saw a variety of new initiatives introduced throughout the association. Many supported the links between science and politics. In the early 1970s, physicist Joel Primack, a recent Stanford Ph.D. who had been appointed to several AAAS committees as a result of the new attention to "young scientists" before and after the Boston meeting, urged the creation of a fellowship program to place scientists in congressional offices. Although the program took several years (and anonymous start-up funding from William Golden) to implement fully, it eventually developed into a multisociety effort coordinated by AAAS. Today, the program is more than twenty-five years old, having placed fellows in a host of congressional offices.[110] Similar fellowship programs in diplomacy and mass media placed scientists in government agencies and media outlets.[111]

Other new initiatives reflected the AAAS commitment to "democratic" participation in science. As in previous eras of change in AAAS history, one marker was the redefinition of what "democracy" meant. In the immediate postwar years, democracy was interpreted in light of the need for open scientific communication; now, democracy meant opening science up to minorities, women, and people with disabilities. Since the early 1950s, AAAS personnel (including especially Wolfle and, after leaving his job at AAAS, Meyerhoff) had been active in collecting data on the problem of "scientific manpower"—how to ensure that enough scientists and engineers were trained to be available for the growing needs of the country.[112] In the late 1960s, as debate about racial tensions appeared throughout American society, AAAS leaders became increasingly concerned about the low numbers of minority scientists. In 1968, the board created a committee to explore ways for minorities to get greater access to scientific education. But the committee soon reported that it could not proceed, for it found deep-seated institutional barriers within AAAS itself—an organization run largely by the elder white males to whom the broad social structures of America had given preference. Without dramatic changes in the participation of young scientists of all races in AAAS affairs, the committee suggested, few changes could be expected. Hearing similar messages from many directions, the board created a Youth Council to bring the voices of younger scientists into its deliberations. A few years later, again following the initiative of women, minorities, and youth, the AAAS in 1972 established both a board committee and an Office of Opportunities in Science to provide full-time professional support for improving the role and participation of minorities and women in science.[113]

In 1974, the board adopted a formal policy committing AAAS to equal opportunity and recognizing "that complex social, economic, and political forces have combined in the past to discourage women, minority, and handicapped persons from entering the sciences and engineering." It urged all professional scientific organizations to "take all measures within [their] power to counteract these historic forces."

Led by Janet Brown, a political scientist, the Office of Opportunities quickly learned to draw on the AAAS's traditional ability to mobilize the resources and coordinate the activities of its affiliated scientific societies. One of the office's first achievements was to secure a Ford Foundation grant to bring together the many projects of the early 1970s to create rosters of women in science.[114] At the same time, the Scientific Manpower Commission that Meyerhoff had once led became a "participating organization" of the AAAS and moved its offices into the AAAS; it would later be renamed the Commission on Professionals in Science and Technology. The two organizations began active collaboration in identifying the number of women and minorities in science, promoting their participation in AAAS affairs, and creating special programs for them. Within a few years, the AAAS was sponsoring symposia, resource guides, workshops, and other activities for women, blacks, Native Americans, and other groups. It also invoked its symbolic powers, passing a resolution in January 1975 formally recognizing "the contributions made by Native Americans in their own traditions of inquiry to the various fields of science, engineering, and medicine," and supporting the development of research programs to study Native American approaches to science.[115]

AAAS leaders had actively supported the contributions of handicapped and disabled scientists in the association's early years. In the 1970s, the association itself began a broad spectrum of programs to ensure access to meetings, laboratories, and other elements of science. The 1976 Boston meeting, held in late February, was the first meeting explicitly made accessible to disabled participants. Interpreters for deaf and Braille programs for blind scientists and engineers were provided, along with wheelchair repair facilities and a special resource room. As a result of its experience, the AAAS prepared a handbook on making meetings accessible to handicapped participants, for which reprint orders were "huge." Members of the AAAS Project on the Handicapped in Science appeared in national media and helped Representative Olin Teague with a panel on research programs for the handicapped.[116]

The association's commitment to issues of equal opportunity continued to expand with the appointment of Shirley Malcom, an African-American science educator who had worked as a program officer at NSF, as director of the Office of Opportunities in Science in 1979; the association was actively seeking out underrepresented groups in science and exploring ways to improve their access both to science as a career and to science as a body of knowledge. In a 1991 memo, Malcom urged scientists to "identify specific locations and activities that provide opportunities for reaching people outside of the audiences that are already attending to information about science. . . . 'Ordinary people' go to the movies, rent videos, participate in sports, and watch television. They also go to the mall, to the grocery store, to the doctor, and to church, and they play Nintendo. . . . We must meet them on these terms."[117]

In 1994, Malcom would be appointed to the National Science Board, the NSF's governing body, in recognition of her pivotal role in improving access to science. Although many AAAS members had served on the NSB, Malcom's appointment marked the first time the honor and responsibility had gone to a AAAS staff member, exemplifying the transformation in AAAS staff to include leaders of the scientific community. Malcom would also be the first AAAS staff member to join the President's Science Advisory Committee. Her achievements symbolized the transformation of the AAAS from a member-driven organization to one driven by its staff.

On the international front, the AAAS had often used its meetings to foster the kind of interaction and exchange that are at the core of its mission to serve scientists. Early in the postwar period, it had occasionally arranged special international sessions, with simultaneous translation and subsequent publication. The first two such meetings (on Soviet Science, 1951, and on Sciences in Communist China, 1960) were at least partially intended to address the political point that science should know no boundaries, even in an era of anti communism. In 1963, a less politically motivated session on Science in Japan was held at the December meeting in Cleveland, funded largely by the NSF.[118] Another innovation in international activity had been the posting of a European correspondent for *Science*, Victor McElheny, in 1964.

But international activities grew substantially in the 1970s, stimulated by 1972 president Glenn Seaborg, a Nobel Laureate in chemistry and former director of the Atomic Energy Commission. Seaborg led the drive to hold the 1973 meeting celebrating AAAS's 125th year in Mexico City. More than five thousand people attended (including Mexican president Luis Echeverria, who had actively supported the meeting and served as honorary chairman), and the meeting led directly to the creation of the Interciencia Association, which helps coordinate the activities of scientific associations throughout the western hemisphere.[119] Leonard Rieser, a Dartmouth researcher and president of the AAAS at the time of the Mexico City meeting, continued to serve as Interciencia's treasurer for many years. The AAAS helped arrange an exchange of scientists between the AAAS and Znaniye (the All-Union Knowledge Society, AAAS's Soviet counterpart). Soon, an ad hoc committee appointed by Seaborg and funded by the Rockefeller Foundation recommended that the AAAS create an International Office, which it did in 1973. Irene Tinker, a political scientist with extensive research experience in planning and urbanization problems in India and Indonesia, was the first director. The new office focused on the kind of interdisciplinary work that the association now found almost routine; its first project was a position paper on the cultural consequences of population change, prepared for a United Nations meeting.

The international programs continued to grow in later years. Under the direction of J. Thomas Ratchford, who would also serve for some years as deputy executive officer, the international office sponsored exchanges with China and other countries, coordinated fact-finding trips, and helped establish sister organizations worldwide. Major AAAS initiatives in international issues often focused on environmental and population issues, with a clear commitment to strengthening scientific work in developing countries; some of the most enduring programs have involved the development of infrastructure and human capacity for scientific research in sub-Saharan Africa. As the cold war ended in the early 1990s, the AAAS worked closely with scientific societies in the former Soviet Union to provide assistance for researchers in those countries.[120] In addition, an increasingly important element of the association's work on scientific freedom and responsibility issues occurred in international contexts, as AAAS members led the drive to free dissident scientists in the Soviet Union during the 1970s and 1980s, and to apply techniques of medical forensics to identify victims of totalitarian regimes in Argentina, Guatemala, Haiti, and elsewhere in the 1980s and 1990s.

Changes in Society

In various small ways, the AAAS, especially in its meetings in the 1970s, reflected the changing standards of American society. For decades, the evening social hours were called "smokers," and were listed by that title in the annual meeting programs. The Biologists' Smoker, in particular, was one of the major events of the meeting. Special notices in the meeting programs reported that "At the Biologists' Smoker and at the AAAS Reception, it may be taken for granted that cigarettes and crackers served *in, or from, a labeled container* were the generous donations of their producers" [emphasis in the original]. But that notice last appeared in the 1966 program. By

1969, social functions were routinely called "receptions" or "social hours." And just two years later, in 1971, the AAAS board adopted a resolution prohibiting smoking at annual meetings.

The AAAS also reflected changes in American society in much more substantial ways. The shift from a predominantly male scientific society to one of inclusion happened quite quickly. Although women had long played a role in AAAS affairs, they had rarely held leadership positions. As late as 1966, the annual meeting program contained a section of "special events for women," organized that year by a committee led by Helen Seaborg, Glenn Seaborg's wife.[121] The reference to women was dropped, but the shift to late-twentieth-century attitudes about women took a few years to overcome inertia; for some years into the 1970s, annual meeting programs included lists of the "welcoming committee," which invariably consisted of the wives of prominent local scientists. Nonetheless, the early 1970s marked the election of the AAAS's first woman president, mathematician Mina Rees. And after twenty years of active AAAS involvement, Margaret Mead finally allowed herself to be nominated for president in 1974—perhaps not coincidentally, in the first year that presidents were elected directly by AAAS members rather than being selected by an elite council (after yet another new constitution had been adopted). Since then, about 25 percent of the presidents and members of the board of directors have been women.

The early women presidents deliberately adopted some changes to tradition. By the 1970s, presidential addresses routinely tackled the complex interplay of science and politics, carrying titles like "The Scientist and the Politician" (Roger Revelle, 1975) and "Science and Its Place in Society" (Emilio Q. Daddario, 1978), although some returned to the old tradition of focusing on the speaker's own field of research. Mead chose the latter route in 1976, exploring the need to reconcile social and natural sciences through a new "human science." But she relied on a multimedia presentation of still photographs, films, and tapes to illustrate her points, making the translation of her spoken talk into a formal published address in *Science* somewhat problematic. She also had an interpreter for the deaf simultaneously translate her talk, "as a way of expressing the concern of the AAAS for the handicapped and my own wish to demonstrate the necessity of always taking into account the multisensory nature of human functioning." In the published version, she also broke with the tradition for presidential addresses and included references to support her arguments.[122]

Several years later, 1983 president Anna Harrison rejected tradition outright. "It has been the custom for *Science* to publish . . . a biography of the incoming president-elect of AAAS," she wrote. "I have bargained instead with the Editor to use this space to explore with you some of my interests and prejudices, hopes and fears, related to science, technology, scientists, society and, of course, AAAS." At a time when the biographical sketches routinely contained information about spouses and children, Harrison said that "the only biographical information that is of significance to you is that I am a physical chemist by training with a modest track record in research and long experience in a liberal arts college that takes science, research, and the development of its students seriously and that I have served the larger community of scientists and society as a member of the National Science Board . . . and president of the American Chemical Society."[123]

Debates about the role that scientists could play in protesting racial discrimination also appeared in AAAS activities. In the mid-1960s, the American Anthropological Association asked the AAAS to appoint a commission on the scientific aspects of race. After Assistant Executive Officer William Kabisch consulted with geneticists, the AAAS declined to endorse such a commission, arguing that no evidence of sufficient quality had been put forward to make such a study worthwhile.[124] A decade later, another race-linked controversy erupted, this time over the selection of AAAS Fellows, that highlighted how ceremonial (and in some ways meaningless) the category of AAAS Fellow had become. Fellows were nominated by the individual disciplinary sections, which judged their qualifications, and approved by the council at the annual meetings. By the mid-1970s, about 16,000 of the association's 114,000 members were Fellows. The nomination process was haphazard: Nobel Laureates, presidential advisors, and other prominent scientists often did not become Fellows until long after they had been honored by others, while relatively minor researchers who happened to work in fields with active nomination committees might become Fellows much earlier in their careers. At the February 1977 meeting in Denver, the controversial educational psychologist Arthur R. Jensen was nominated by the psychology section for his career of research contributions. But one of those papers had been a 1969 article claiming that blacks were, for genetic reasons, less intelligent than whites. At the council meeting, a group of black scientists vigorously objected to Jensen's nomination, but found itself stymied by arcane and poorly applied voting rules. They stalked out in disgust, and Harvard University health administrator William D. Wallace announced to waiting reporters that he would resign from the AAAS. Margaret Mead, by then a former chairman of the AAAS and a revered figure in American science, followed, and also denounced Jensen. But she refused to resign, telling reporters, "I never resign from anything. I'd rather stay in and fight from within."[125]

The Jensen incident led the AAAS to suspend the naming of fellows. After thirty thousand members voted 2–1 in support of retaining the fellow category, however, the board refined the selection process to allow for reconsideration and investigation of nominations. Today, several hundred Fellows are named each year; under 10 percent of the overall AAAS membership are Fellows.[126]

The AAAS commitment to equal opportunity expressed itself in other ways as well. In 1975, the council adopted a resolution deploring "any form of discrimination on the basis of sexual behavior between consenting adults in private." In 1978, the board moved the scheduled 1979 Chicago meeting to Houston on just ten months' notice, and voted to require that future annual meetings occur only in states that had ratified the Equal Rights Amendment. (By its actions, the AAAS joined other professional societies including the American Psychological Association, the Modern Language Association, and the National Education Association.) Similarly, in 1993, after voters in Colorado approved a referendum limiting the rights of homosexuals, the AAAS canceled plans for a 1999 meeting in Denver. Although the referendum was subsequently overturned in court, the 1999 meeting was held in Anaheim, California; at this writing, the 2003 meeting is scheduled for Denver.[127]

Not everything in AAAS history was serious or weighty. In a 1972 *AAAS Bulletin*, the editors reported receiving two disconcerting letters in recent months: one addressed to the American Association for Advertising Science and one to the American Association for the Replacement of Science.[128]

New Needs, New Structures

As the AAAS struggled both with the ordinary demands of an organization for growth and with defining its own role in American science, it briefly considered dramatically expanding the membership pool on which it would draw. In his presidential letter, 1970 president Athelstan Spilhaus noted that the board had gone on record "as favoring in principle an order-of-magnitude membership increase in the next decade." Such an increase, which would have brought membership from 130,000 to over a million, "means that we must actively serve the young group right down to school age; it also means that in promoting the public understanding of science, we will perhaps encourage nonscientists to become members." The goal of attracting nonscientists reflected a deliberate shift away from the inward focus of advancing science by serving scientists' special needs. "If we are to move more deeply into science and public affairs," wrote Spilhaus, "we must have within our council people from disciplines other than science who can be effective in implementing science in the promotion of human welfare."[129] In the end, the AAAS realized that such a broad expansion of its membership would dilute its special ability to represent the scientific community. But the need to open up AAAS governance to broader constituencies was clear, and the die was cast for a major reorganization and democratization of AAAS affairs.[130]

New Forms of Democracy

Although the AAAS had been founded as an organization of individuals, it had by the 1920s become a creature of its affiliate groups. Indeed, in 1917, James McKeen Cattell had proposed that the AAAS be officially reconstituted as an association of societies, not individuals. The affiliates had the dominant voice on the AAAS council, and through it on the executive committee that supervised the daily operation of the AAAS. But as the specialized societies grew less involved in the AAAS through the 1950s and 1960s, individual members chafed more and more at their inability to participate fully in AAAS governance. Some reforms had been achieved, including the shift from an executive committee to a board of directors in the 1950s, but many tensions remained. The recurring organizational frustration of having both individual members and nearly three hundred affiliated organizations had flared up in the early 1960s, and a new committee on council affairs wrestled with the problem. As Dael Wolfle relates in his history of the period, a series of attempts to address the composition of the council and the election of officers occurred during the 1960s. The discussions were again tied up with problems of "democracy"—this time, the question was whether a "representative" democracy was as "democratic" as a "participatory" democracy. The discussions also occurred in the context of the practical limitations of a society that held each annual meeting in a different city, and thus could not count on recurring attendance by the same members from year to year.

The discussions also, of course, reflected a difference of opinion about whether the AAAS represented all scientists as individuals, or science as a whole. "We are less a scientific community," wrote the political scientist Don Price in his 1967 president's letter, "than a collection of scientific communities."[131] One frequent comparison—a recurring one throughout the century—was to the National Academy of Sciences. Many AAAS members were proud that they represented the broad base of scientific activity throughout the country, not merely the scientific elite. In an American society committed to the ideal that the best decisions are based on broad, open debate and discussion, AAAS members were frustrated that the NAS, by virtue of its congressional charter, commanded more respect. Yet at the same time, they recognized that scientific knowledge itself was not the product of a democratic process (though of one in which open, informed debate was at least in the ideal case demanded). The leadership of the AAAS frequently consisted of the prominent scientists who were equally likely to be members of the NAS or its companion groups, the National Academy of Engineering and the Institute of Medicine—chemist Henry Eyring (1965), oceanographer Roger Revelle (1974), statistician Frederick Mosteller (1981), engineer Sheila Widnall (1988), physicist Walter Massey (1989), and Nobelists Leon Lederman (1992) and F. Sherwood Rowland (1993), to name just a few.

Eventually, these philosophical debates fell before the practical frustrations with the unwieldy council, and especially its dominance by affiliated organizations that did not, in actual practice, contribute significantly to the ongoing activities of the association. In the early 1970s, a series of changes were introduced that made the council a representative largely of the sections, which were themselves representative of the individual members. In addition, the major officers (president-elect, nominating committee, and board of directors) were elected by the individual members, rather than by the council. Four principles were embodied in the new constitution:

1. Members should have the opportunity to participate more widely in the association's governance.
2. The general policies should be determined by a council, which both represented the membership and was of a size that allowed it to be an effective legislative body.
3. The board of directors should be the executive arm of the association.
4. The constitution itself should be lean, and subject to amendment by the members. The bylaws could be more detailed, and subject to change by vote of the council.

With those changes in the 1970s, the AAAS completed a circle. Created by individuals out of an earlier organization of geologists and naturalists, through the next century and a half it spawned new organizations and drew its strength from its continuing relationship with those organizations; now it returned to being an association of individuals committed to a broad vision of science that necessarily crossed the many disciplinary, methodological, and organizational boundaries inherent in science.

One effect of the change in organization, however, was to highlight the degree to which many AAAS members joined solely to receive the journal *Science*. In the years just prior to the new constitution, typically 90 percent of members enrolled in a section. However, since the adoption of the new constitution, when members have had to declare a primary section for the purposes of voting, almost half have failed to do so. Some sections retained large portions of their members, but several of the large sections (such as chemistry, biological sciences, and medical sciences) lost half or more of their members.[132]

The Annual Meeting Adapts

The 1970s also saw other significant changes from earlier periods in AAAS history. Throughout the 1960s, fewer affiliated societies held separate programs either as part of or alongside the AAAS meeting. In 1972, the last of the post-Christmas Convocation Week meetings, only twenty-one other groups had their own meeting schedules published in the AAAS program book, ranging from a full multiday program arranged by the American Society of Zoologists through a single session organized by the National Speleological Society. When the annual meetings resumed in 1974 (after the special international meeting in 1973 in Mexico City celebrating the AAAS's 125th year), program planning had been turned over entirely to the AAAS sections and the program committee, which selected the major themes around which the meeting would be organized. *Science* editor Abelson noted, "Only a minority of scientists are qualified by temperament and interest to grapple with social problems. However, a substantial fraction of those who are so inclined are members of AAAS, and many of them attend and participate in the meetings." While many members found that the meetings organized to meet their interests were exciting and challenging, others continued to complain both about the lack of real scientific "news" at the meetings, and at the uneven quality of sessions, the result of differing standards and expectations in different areas of science. In 1977, Executive Officer William Carey echoed concerns expressed by Meyerhoff twenty-five years earlier: the meeting "has evolved as a process of interdisciplinary communication . . . and an instrument for public understanding. But this focus on the horizontal access can be carried too far, at the expense of straightforward discussion of main directions in scientific discovery."[133] The tension between scientific disciplines and multiple definitions of what "interdisciplinary" might mean continued.

Despite these concerns, the annual meetings continued to be an eclectic mix of technical sessions and interdisciplinary workshops devoted to issues at the interface of science and society. Many of the new programs on science and ethics, human welfare, and government funding emerging at AAAS headquarters–sponsored symposia at the meetings. Plenary lectures (which had replaced the vice presidential lectures of earlier years) reflected the evolving concerns of the scientific community: in 1972, biologist René Dubos spoke on "Humanizing the Earth," while in 1977, physician Lewis Thomas spoke on "Biomedical Science and Human Health: The Long Range Prospects." But the thrill of science was also part of each meeting; the 1981 meeting, for example, featured astronomy, with Philip Morrison of MIT speaking on "The Rude Law of the Frontier: Deep Sky Astronomy Today," and Freeman Dyson of Princeton speaking on "Infinite in All Directions."[134]

The interplay between science and politics engaged many meetings. Lectures and symposia on the problems of nuclear war, for example, were common throughout the 1970s and 1980s. In 1978, the U.S. Senate and House subcommittees with principal jurisdiction over science and technology policy held joint hearings at the annual meeting. Although some presidents declined invitations to address the AAAS, Harry Truman and George Bush accepted, as did vice presidents Nelson D. Rockefeller and Albert Gore. In 1998, as the association celebrated its 150th year, President William Jefferson Clinton came to the Philadelphia meeting to offer his birthday greetings, using the occasion to announce several personnel changes. Reflecting the prominence of the AAAS at the end of the century, every one of the people he named was active in the association: former AAAS board member John H. Gibbons was retiring as science advisor, AAAS Fellow Neal Lane was named to succeed him, and former AAAS president Rita Colwell was named to replace Lane as director of the National Science Foundation.

The meetings also served the public outreach function, especially after the board committed AAAS internal funding to support a Public Science Day to lead into meetings, beginning in 1989. Each year, thousands of school-children in the meeting city attend lectures and workshops on the day before the meeting begins. Many of these schoolchildren come from the inner cities, reflecting the current AAAS perspective on democracy, in which the association is committed to making science available throughout society. At the 1996 Atlanta meeting, Public Science Day for the first time included telecommunication links beyond the meeting city (that year, throughout the state of Georgia), and Public Science Day at the 1998 Philadelphia meeting featured links to schools in England as well as ties to ongoing school science projects.

In 1991, responding to the concerns over quality and focus, the AAAS moved to a single program committee, with peer-reviewed sessions. While the process of soliciting and promoting sessions would still take place at the section (which is to say, scientific discipline or specialty area) level, the single program committee in principle would attempt to create a more even distribution of topics and quality across the meeting. It took several years for the system to be fully implemented—one meeting director was let go in the midst of the annual semichaos of arranging a meeting—but that is the system the AAAS operates under at this writing.

New Leadership, New Initiatives

Throughout these dramatic changes in official AAAS structure and emphases, the staff continued to develop expertise in creating and managing diverse programs. But the staff initiatives were outstripping the association's ability to manage them. In the early 1970s, the dramatic increases in national inflation, combined with unexpected drops in membership and advertising revenue, led to severe shortfalls in AAAS budgets. The costs of the 1973 Mexico City meeting also put a strain on the association's finances. Programs and people were cut, including the number of pages published by *Science* and the entire *AAAS Bulletin* (which was replaced by a "AAAS News" section that appeared monthly in *Science*). The financial straits continued, however, and in mid-1974, executive officer William Bevan abruptly left the AAAS. *Science* editor Philip Abelson served as interim executive officer, and many staff members credited him with maintaining morale while making the financial adjustments needed to keep the association solvent.[135]

At the beginning of 1975, William D. Carey became executive officer. Like Abelson and Treasurer William Golden, Carey had been a key player in postwar science politics, having been at the U.S. Bureau of the Budget during the years after the war when federal funding mechanisms for science were being developed. Though Carey was familiar with many sciences, he had no formal scientific training. For him, the attraction of the AAAS was the opportunity "to try to define more clearly where science and technology belong in the country's planning and priorities."[136] Carey could be charming, making him extremely effective in various settings. Speaking to the American Library Association, he began by admitting "a great will to say inspiring things on a mellow Sunday." But, he said, "I soon got over it because in consulting Volume V of the Oxford English Dictionary I discovered that a 'keynote' is the lowest note of the scale. If I hadn't been worried already by the prospect of facing so many librarians I wouldn't have troubled to look it up, but there it is. Knowledge isn't always friendly." To other audiences, he could be blunt. A letter to the secretary of defense began *in medias res*: "I must tell you that the otherwise excellent brochure on *Soviet Military Power* went off the rails badly, in my opinion, in contending that U.S.-sponsored scientific exchanges and scientific communication practices enhance Soviet military power."[137]

Reflecting Carey's experience with federal budgets, one of his first new initiatives was an attempt in 1976 to analyze the place of science in the budget. The study was conducted by Willis Shapley, a former senior colleague of Carey's at the Bureau of the Budget who had also worked at NASA, and "the results . . . provide us, at last, with a perceptive and critical treatment of the process for allocating resources to research and development," Carey wrote.[138] Sponsored by the committee on science and public policy and funded by the Alfred P. Sloan Foundation, the R&D reports quickly became one of the AAAS's most visible and influential products, and have now passed their twentieth anniversary. An important aspect of their effectiveness was that, after the first few years, the integrity of the report was sufficiently respected that cooperation from Congress and the various agency offices was easier to get. Moreover, in a sign of the AAAS's continuing success at brokering activities that have drawn on many different organizations in the scientific community, other professional societies soon contributed significantly to the analysis of portions of the budget. By the mid-1980s, the contribution of the other societies was so great that the annual volumes no longer carried individual author credits, but were attributed to the Intersociety Working Group (today, more than twenty organizations) that continues to coordinate the series under the direction of AAAS staff member Al Teich, a political scientist.

Under Carey's guidance, the AAAS also launched a major new initiative in public understanding of science when *Science*'s "Research News" editor, Allen Hammond, proposed that the association create a glossy magazine that would be somewhere between *Scientific American* and *Popular Science* in tone and level. Like many of the newer AAAS initiatives, this one came from the staff rather than the membership. The AAAS board approved of the plans, and the magazine (called *Science 78* in its earliest incarnation, and launched as *Science 80*) soon became a major AAAS activity.[139] The AAAS board clearly recognized that the decision about *Science 8X* (as the magazine was sometimes called within the AAAS's offices) was not "an isolated one of simply introducing new product lines; it goes farther, and involves the larger goals and resources of the AAAS." Discussions at the board explored the goals of public understanding, often reiterating the positions developed in the late

1960s: the AAAS was not talking down to the public, but rather was trying "to communicate across a barrier between the scientific community and the public."[140] The new magazine was a lineal descendant of the new impetus for public communication of science that began after World War II. But while the AAAS's formally stated goals remained constant, the rationale behind those goals had evolved to take into account new ideas about the relationship between science and society. That evolution is most notable in a comparison of the fate of the proposed 1963 magazine on *Science and Public Policy*, which had died when it dared to suggest that scientists did not have the last word on public issues, with the triumphant rise of *Science 80*, committed only to presenting "balanced views." While the early activities of the AAAS in the postwar period were motivated by a firm belief in the moral superiority of science, the later ones reflected a different, more humble attitude toward the place of science in society.

Science 8X was an immediate editorial success. It rode the "science boom" of the early 1980s, fueled by the rapid growth of personal computer advertising and by the sense among many people in the publishing industry that "science" was a new hot area to pursue. For several years in a row, the magazine won major national awards for its content and design. But when the advertising market collapsed in 1984, the AAAS discovered that the new magazine could also be a severe financial liability. In 1986, when Time, Incorporated, offered to buy *Science 8X* (in order to add its subscribers to Time's own *Discover*), the board reluctantly agreed.[141]

Science itself continued to flourish under Abelson's leadership. In 1980, it published a report by Luis Alvarez (a Nobel Laureate in physics) and others claiming that a meteor crash had caused the extinction of the dinosaurs 165 million years ago. The paper led to years of controversy, mitigated only slightly by the discovery of the Chicxulub meteor crater in Mexico's Yucatan peninsula. Two years later, "Research News" reporter Jean Marx's article on AIDS provided one of the first comprehensive looks at the emerging epidemic.[142]

A New Objective

In the 1970s and 1980s, the AAAS's longtime concern with issues of scientific freedom and human rights also took on new life with the growth of conflicts between ideals of scientific practice and the realities of a scientific world largely funded by industries and governments, especially in emerging areas of science like genetic engineering. In 1970, the board established a committee on Scientific Freedom and Responsibility after it was asked by Senator Edmund Muskie to investigate whether two Atomic Energy Commission scientists had been harassed by their superiors. During the committee's several years of deliberations, the AAAS itself contributed to the dialogue about scientific responsibility by publishing, in *Science*, the famous "Berg et al." letter calling for a moratorium on experiments with recombinant DNA. The committee was led by Allen Astin, director emeritus of the National Bureau of Standards; nearly twenty years earlier, political pressures had forced Astin to briefly resign from NBS when he refused to rescind a report sharply critical of an unproven battery additive. After several years of work, the committee issued a landmark report on *Scientific Freedom and Responsibility*, written by Harvard biologist John T. Edsall. The report reflected the maturity of the scientists whose careers had blossomed in the postwar years as well as the effects of the political protests of the 1960s and the direct criticism of the AAAS at its most recent annual meetings. Now the scientific community recognized the breadth and interconnection of a host of issues; the report included sections on communication and theft of scientific ideas; principles of informed consent; the conflict between science and secrecy; and the multiple and complex effects of technology and innovation.

The report's impact was based, in part, on its willingness to take unequivocal stands. For example, the report noted that the committee "accepts the advancement of knowledge as a fundamental value; yet certain means for the achievement of knowledge must be either renounced or employed only with special safeguards."[143] Then it condemned the lack of informed consent in medical research, calling attention especially to the misuse of psychology as a tool for torture in dictatorships worldwide. (The infamous Tuskegee syphilis experiments had also recently been made public.) The report also expressed much of the conflicted feeling of scientists in the early 1970s: "There is an enduring need for the devoted researcher-teacher, who loves science for its inherent beauty and fascination and seeks to impart that love and understanding to others. Yet increasingly, for many of us, it is impossible to feel the same delighted fascination with science that we once did, without also being deeply concerned with the uses and misuses of science that will largely determine the future of mankind."[144] The report attempted to identify procedures for resolving conflicts, and called on the professional scientific societies (which, it acknowledged, the AAAS could help coordinate through the affiliate structure) to serve as appeals bodies and to develop codes of ethics.

Recognizing the importance of the constellation of issues identified by the report, the board made the committee permanent and soon added the words "to foster scientific freedom and responsibility" to the AAAS's constitutional objectives. Members of the committee have been extremely active in the ensuing two decades, protesting the restrictions imposed on Soviet scientists, and investigating various conflicts between scientists, corporations, universities, and other institutions in the United States. The AAAS staff became deeply involved in this area, creating programs on research integrity and scientific misconduct, the effects of national security controls on unclassified research, and a host of professional ethics issues.

Perhaps the AAAS's best-known activity in this area, beginning in 1984, was to create teams of forensic anthropologists to help investigate human rights abuses in Argentina, the Philippines, Haiti, and elsewhere. Led by Clyde Snow, a Texas-bred, hard-drinking freelance consultant from Oklahoma who had become one of the world's leading experts at using anthropological techniques to identify the victims of accidents and crimes, the AAAS teams both identified specific victims and conducted workshops in countries that had suffered through murderous dictatorships to train physicians, lawyers, archaeologists, and anthropologists in how to assemble evidence of crimes. Often coordinated by staff at AAAS headquarters, the teams showed how scientific evidence could address some of the most morally repellant human rights abuses of the recent past.[145]

New Home, New Editor

Carey's tenure also included more prosaic achievements. At the end of World War II, the AAAS had still been run out of a few dank and dark rooms at the Smithsonian Institution. Later, as it took over production of *Science* and began to build programs in the late 1940s, it moved to a series of houses at Scott Circle in downtown Washington. In 1954, it tore down the houses and built a more permanent home, a distinctive building at 1515 Massachusetts Avenue, with special louvered windows and energy-efficient construction. By the early 1980s, however, the building was bursting at the seams. Carey's leadership and the professional staff's ability to acquire grant funding—both from government agencies and from private philanthropy—had led to significant expansions of programs in the areas of science and public policy, public understanding of science, and scientific freedom and responsibility. Staff lists in the late 1960s and early 1970s rarely showed more than twenty employees. The expansion of the Bevan years brought the staff to around fifty. By 1980, the senior staff alone totaled more than fifty. The continuing success and growth of *Science* and the creation of *Science 8X* had also created new needs, and the AAAS was renting rooms in a variety of local office buildings. When one prospective employee interviewed with *Science* managing editor Robert Ormes in 1980, he was told that even if a job were available, he could not be hired because there was no place to put him—even the corridors already had all the desks they could handle. In 1985, the AAAS took a long-term lease on the top half of a new building at 1333 H Street in downtown Washington.

The year before had seen another transition in leadership, from Philip Abelson to Daniel Koshland as editor of *Science*. A prominent biochemist at the University of California at Berkeley and the chairman of the editorial board of the *Proceedings of the National Academy of Sciences*, Koshland was already sixty-five when he took the reins at *Science*, but continued the example set by his predecessor of maintaining an active laboratory. Moreover, instead of commuting across town, he commuted back to Berkeley. In the ten years he served as editor of *Science*, he also published nearly one hundred peer-reviewed papers.[146]

Koshland oversaw major changes in *Science*, typified by new typography and format for the journal. He acknowledged that:

> A design touch, the dot over the capital I, may elicit comment. To some it will represent the height of modernism, a sign that *Science* is becoming avant-garde and moving with the new era. To others, it will be the triumph of the typographical error over the forces of scholarship. To me, it represents a balloon rising above earthbound reality from which to look forevermore for distant intellectual horizons. It represents the light at the end of the tunnel, the globe whose environment we must study and protect, the hole in the argument that must be plugged. It is the beginning of the Big Bang, the first wheel, the peephole into the future, the period at the end of Q.E.D. It symbolizes imagination and the willingness of scientists to battle conformism, for these lie at the heart of all great science.[147]

Koshland reorganized the journal, added the "This Week in *Science*" quick summary to lead off each issue, and introduced various news sections and special issues. Together, these changes helped *Science* maintain its position as one of the two leading international journals in science. His decade of service also marked a time of tremendous growth in the overall scientific enterprise and in the mechanics of technical publishing, leading him to introduce a series of innovations that kept the journal from losing its edge. Perhaps the most important was the board of reviewing editors, which provided a first pass at all submitted manuscripts. This process let the editors cull the most promising from the six thousand or more submitted each year before sending them out for more complete review.

During the Koshland years, the journal's finances also underwent a major transition. After nearly fifty years of letting an outside firm sell the journal's advertising space, the AAAS brought the function back in-house, expecting to take at least several years to recover the costs associated with that move. Instead, under the leadership of associate publisher Beth Rosner, within two years increased revenues from advertising were flowing back to the journal. The new money allowed *Science* to publish more articles, and to expand its international efforts, both by inviting submissions and by providing news coverage from around the world.[148]

Koshland's editorial leadership was not without controversy, especially in terms of the relationship between the journal and the AAAS. Some members objected to "joining" AAAS (and thus subsidizing its activities), when all they really wanted was a subscription to *Science*.[149] For others, the objection was the opposite: *Science* did not accurately reflect the membership of the AAAS. Despite the efforts of Abelson to expand the coverage of physical sciences, the journal had for many decades been perceived as being primarily a location for publishing in the biological sciences. Under Koshland, some AAAS members felt that the balance had tipped even further away from the physical and, especially, social sciences. Koshland responded, in presentations at section meetings during the AAAS annual meetings, that he could only publish what he received, when it met *Science*'s high standards. Moreover, he pointed out, many of the issues of concern to members were covered by the expanded news sections that he had introduced. No major flare-ups resulted from the disputes, but they illustrate the continuing tensions between a large association and a journal that it both controls and depends on for revenue.[150]

Overall, in the Koshland years *Science* clearly continued to serve as the AAAS's main organ for "advancing" science. The number of pages devoted to original research reports rose from 1500 pages in 1985 to 2,300 in 1994, and the journal remained "the" location for publishing cutting-edge research in such fields as immunology and genetics. In 1985, *Science* published the first major article on the polymerase chain reaction (PCR) technique that revolutionized research on DNA; four years later, it described the first human gene discovered with the new techniques, for cystic fibrosis. In the early 1990s, the staff was growing at a rate of nearly 20 percent each year, and many more of the staff had significant research experience of their own, with more Ph.D.s on staff than ever before.[151] According to the Institute for Scientific Information, *Science* in 1996 was the fourth most-cited journal in the world and is in the top ten for both impact and immediacy.[152]

The Fin-de-Siècle Generation

The middle 1980s were eventful years in AAAS history, with the move into a new building, the appointment of Daniel Koshland as editor of *Science*, and the closing for financial reasons of AAAS's venture in popular science publishing, *Science 8X*. They also marked the passing from the scene of many from the generation of AAAS leaders who had experienced World War II and its immediate effect on the scientific enterprise. Although Koshland had worked on the Manhattan Project, which developed the atomic bomb, and treasurer William Golden had been part of the original Atomic Energy Commission staff, virtually all of the senior AAAS and *Science* staff now largely responsible for AAAS affairs were younger. Their entire educational and professional careers had been the product of the federal largesse that shaped science in the second half of the twentieth century. The AAAS was no longer an organization in which a cadre of individuals ran programs out of their pockets, meeting annually for mutual reinforcement. Instead, it had become a scientific institution in its own right, a group with its own buildings, identity, and self-sustaining programs. Many AAAS staff members had spent their careers in organizations like the AAAS, the National Science Foundation, or other professional societies. While they might have some laboratory or teaching experience, those traditional activities did not define their careers in the same way that they had defined the careers of earlier AAAS staff members.[153]

The institutional nature of the AAAS became clear at the end of the decade, when Alvin Trivelpiece, a U.S. Department of Energy official, took over in 1987 as executive officer of the AAAS on William Carey's retirement. Author of one of the standard textbooks in plasma physics, Trivelpiece came highly recommended, and with an extensive background in both government and management. As part of his initial efforts at the AAAS, he began a review of the many "offices" that had sprung up during the Bevan and Carey years to handle the expansion of AAAS activities—offices of communications, international science, opportunities in science, science policy issues, and so on.

Less than two years later, Trivelpiece was gone, having become director of the Oak Ridge National Laboratory. But the institution continued on its path, without the wrenching dislocations or uncertainties that had occurred in 1953 and 1974 when other executive officers had suddenly left. (The 1970 transition from Wolfle to Bevan had been expected and planned for well in advance.) The new executive officer, Richard Nicholson, continued the review inaugurated by Trivelpiece, and soon reorganized AAAS activities into three directorates: Education and Human Resources, Science and Policy, and International Programs.

The choice of Nicholson also continued the postwar tradition of selecting AAAS administrative leaders from those with extensive government experience rather than laboratory time. Although his own doctoral work in chemistry had been highly cited, he spent most of his career at the National Science Foundation in increasingly responsible roles, rising to assistant director for mathematics and physical sciences (a presidential appointment) before joining the AAAS. At the time of his appointment some people worried that he did not have the public presence of his predecessors (especially Carey, who could be a highly engaging public speaker), but others lauded his professionalism and tact.[154]

One of Nicholson's continuing important priorities was education. The restructuring of 1990 put outside the directorate structure—and thus reporting directly to Nicholson—one major program: Project 2061. The project emerged in the early 1980s as the AAAS struggled to find a way to convert individual education projects into a coherent long-range program. The association's senior education officer, F. James Rutherford, had been a key participant in Project Physics, one of the science curriculum development projects of the 1960s that paralleled the AAAS's own Science—A Process Approach. He had joined the AAAS after two presidential appointments, one as assistant director of the National Science Foundation and one as assistant secretary of the newly created U.S. Department of Education. Led by Rutherford, Project 2061 developed a rationale for science education (published as *Science for All Americans* in 1989) and then a series of *Benchmarks for Science Literacy* (1993) that would help guide teachers, curriculum developers, and others trying to create specific educational materials. Initially funded by the Carnegie Corporation of New York, Project 2061 drew on the AAAS's strengths as a coordinator of other organizations and as a membership body with wide representation in the sciences. Later, funding came from a consortium of foundations and government agencies, including several state educational offices. As with Science—A Process Approach, Project 2061 stressed the value of interactive learning and the importance of teaching the process of science rather than specific scientific facts or theories.[155]

Unlike many of the earlier AAAS education initiatives, Project 2061 has left a clear trail of major publications and reports. Yet, also as with so many other AAAS projects, its impact is difficult to pinpoint and has come primarily through its role as a catalyst in the discussions of others. Most notably, a draft version of *Benchmarks* was used by the National Academy of Sciences in 1992 as it developed the National Science Education Standards published in 1995. To many observers, the national standards are clearly based on the AAAS's benchmarks. Thus Project 2061 shaped both national- and state-level debates about science education. Like other AAAS activities, it may have sown the seeds of developments that were eventually carried out by other organizations.

The AAAS did not entirely cease to be a creature of its members and elected leaders in the 1980s and 1990s. Leading scientists like F. Sherwood Rowland, who won a Nobel Prize for his work on ozone depletion, and Jane Lubchenco, an internationally known ecologist, served as AAAS presidents and used their presidential platforms to urge new attention to global problems. But with the daily projects that defined the AAAS now almost exclusively run by the professional staff with funds raised by grants, and with *Science* operating largely independently of the association, the members served largely as representatives of contemporary concerns. In that way, they continued the important role of the AAAS as an intellectual forum, a place for introducing and debating issues that would be applied in specific instances far from the AAAS itself. As in the early years of the AAAS, the board, the council, or their committees issued resolutions addressed to legislatures and to the world at large. Some deplored genocide in Cambodia and violations of human rights in Nicaragua, while others urged the U.S. president to reestablish a Science Advisory Committee or supported specific pieces of legislation such as the Endangered Species Act and metric education. Many resolutions were tied to the concerns of committees like the Committee on Scientific Freedom and Responsibility, such as one that deplored the censorship of *Science* when it reached the Soviet Union (especially when the journal carried news stories featuring the persecution of Soviet scientists). Committees and sections sponsored programs at the annual meetings that served as forums for discussing new issues, and served as advisors to staff-driven projects such as Public Science Day or the Congressional and Mass Media Fellows programs. And members contributed in another direct way: although grant funding supports many AAAS activities, the basic operating costs of the AAAS and many of its new initiatives are supported by member dues.

Those new activities continue to emerge. In 1991, the AAAS announced the creation of the *Online Journal of Current Clinical Trials*, the world's first fully electronic journal. Though it recognized that the format and funding for online publications was still unclear, the AAAS also believed that the challenges to the scientific community were sufficiently important that one of its responsibilities would be to explore those challenges. The online journal was sold a few years later, but the experience gained contributed both to the general growth of online publishing and to *Science*'s own online presence beginning in 1995, which today includes "*Science*'s Next Wave" (a series of features exploring topics of interest to working scientists), "*Science*Now" (fast-breaking news in science), and "EurekAlert!" (a service for science journalists, providing early access to new developments in science). In public understanding of science, the AAAS received NSF funding in the early 1990s to create *Kinetic City Super Crew*, a radio show targeted at children and distributed over the emerging network of radio stations aimed at children. The show soon won a series of awards, including a Peabody, one of the most prestigious in broadcasting. It, too, led to online initiatives like the Kinetic City Super Crew web site and ScienceNetLinks, a website that rates the quality of other science websites. Reflecting the long-standing concern of the AAAS for government and policy issues, in 1994 the association established a Center for Science, Technology, and Congress. Two years later, it created the Program of Dialogue Between Science and Religion, chaired by former AAAS president Francisco Ayala and supported by financier John Templeton.

Two of the most recent developments demonstrate the continuing ability of the association to adapt to change. In 1995, neuroscientist Floyd Bloom took on the reins of *Science* from Koshland, continuing the tradition of putting highly successful laboratory scientists into the chair of one of the world's leading scientific journals. Just a year after taking over, Bloom oversaw the rapid publication of new research claiming to show the presence of signs of previous life in meteorites that had originated on Mars. Bloom also continued the expansion of *Science* into cyberspace. As this book was in press, Bloom announced that he would step down after a single five-year term as editor; a new editor will lead *Science* into the twenty-first century. The second event with a major impact on the association was the construction of a new building. In 1996, the association moved into 1200 New York Avenue in downtown Washington, a building that continues the tradition of technologically advanced AAAS homes. Named for treasurer and major AAAS benefactor William T. Golden, the new building includes a model classroom, an exhibit area, and a small conference center. AAAS leaders expect that the new space will lead to new initiatives in which the AAAS becomes a physical as well as intellectual space. The building cost about $65 million (including the price of the land); symbolizing the complex relationship that the AAAS now has with other institutions, much of the construction cost came from tax-free bonds issued by the District of Columbia government. The most noticeable technological advance is a computer-operated elevator system that directs users to specific elevator cars, allowing it to manage the elevator load by sending cars only to certain floors on each run. Less obvious to casual observers, but equally important, are environmental control systems that use no artificial coolants and a light-sensor system that adjusts the artificial lights based on the amount of natural light coming in through the windows.[156]

Conclusion

Today, with the AAAS having celebrated its 150th anniversary and looking forward to the twenty-first century, it is too soon to tell which of its current programs will thrive and which will give way to new initiatives. Some, like the *Online Journal of Current Clinical Trials*, may be important but ahead of their time. Others may be the harbingers of successes of the future; many people point with special pride to the international activities of the Committee on Scientific Freedom and Responsibility, to the annual analyses of the federal R&D budget, and to the growing reach of Public Science Day as programs whereby the AAAS has become the home for ongoing programs with clear impact on the national and international scientific community. The organization itself is vibrant, with about 145,000 members, a highly successful journal that will soon celebrate its own 120th birthday, and a new building that continues the AAAS tradition of including state-of-the-art technology.

During the postwar years, the AAAS moved from being a small organization defined solely by its leaders to an established institution with widely varying programs, defined as much by the programs as by the people leading the programs. As science and technology themselves have grown and occupied increasingly fundamental roles in a vast array of modern institutions, so AAAS programs and interests have expanded to reflect the diversity of science. In that way, the AAAS continues to represent the "democratic" instincts of its founders, altering its conception of democracy to account for changing times.

The defining struggle at the AAAS in the years just after World War II was over the idea of "interdisciplinarity" and its implications for AAAS activities. The issue remains central today. In December 1997, as the AAAS entered its 150th year, the board issued a call for "conversations" on the future of the AAAS—and immediately drew fire for its attention to interdisciplinary science. In the position paper that opened the conversations, AAAS board members wrote that each issue identified as urgent for AAAS and for science more generally,

> from population and the environment to the public's understanding of science, seemed to have radically outgrown its previously accepted conceptual framing. For each of these issues, new theories, explanations, and cause-effect relationships were appearing on the horizon. . . . There has been a movement away from assigning simple causes to complex, physical, biological, and social phenomena. Feedbacks and synergies are now known to complicate causal stories that once were regarded as simple and linear. Dynamic cross-systemic explanations are sought where static and reductionist models once prevailed.[157]

In the online debate encouraged by the position paper, Paul Gross, a former director of the Woods Hole Oceanographic Institute and coauthor of *Higher Superstition: The Academic Left and Its Quarrel with Science*, wrote back sharply: "Nobody can disagree with the broad SENTIMENTS of the position paper. Only the specifics—the assertions and strong hints—will trouble anyone who actually READS them. There is bombast, for example, about causality, about a "paradigm shift" in science. . . . No such shift is identified, however, nor has one taken place recently in any important body of science I know of."[158]

Like the editorial published by Howard Meyerhoff shortly before he resigned in 1953, Gross's electronic posting represented a commitment to the long-standing traditions of the individual scientific disciplines—the disciplines which, through their section memberships, defined the membership of the AAAS. The board's statement also represented a long-standing AAAS tradition—that of prominent scientists seeking to lead members to a new vision of science, one that escaped the confines of individual disciplines without losing the intellectual rigor that disciplines provided.

This history is written just as the new AAAS conversations begin, and so cannot predict where those conversations will lead. But the opening contributions clearly show that the issues identified by all three chapters of this history—issues of membership, representation, disciplinary links, and public outreach—will continue to define the AAAS into the twenty-first century. The enduring strengths of the AAAS emerge from its origins: its wide representation of American science (as opposed to the elite representation of organizations like the National Academy of Sciences); its commitment to geographic diversity in its activities (as demonstrated by its regional divisions and its peripatetic annual meeting); its role as a site for scientific presentation and publication; and its continuing commitment to engagement with the public beyond the scientific community. New perspectives on these commitments have emerged in the half century covered by this chapter, particularly the AAAS's commitment to international programs, to issues of human diversity, and to science education at all levels. But a stable central core defines the association, and will likely do so well into the next century: the AAAS is a forum in which the scientific community can define itself, coordinate its activities, and hone the advocacy and work by which the advancement of science can be assured.

APPENDIX A:
THE ARDEN HOUSE STATEMENT

In 1951, the board of directors (then called the executive committee) met with several consultants to reassess the function of the AAAS. The statement prepared at that meeting has served as a guide to officers and staff ever since. Because the meeting was held in Arden House, in Harriman, New York, the statement has been familiarly known as the Arden House Statement.

Whatever its obligations to other groups and whatever its opportunities in wider fields, the AAAS is an organization of scientists for science. The AAAS must, first of all, serve scientists and science in such a way as to command the confidence and backing of the scientists of this country. Otherwise it will be in no position to meet its wider opportunities.

This central principle indicates the necessity that the AAAS reexamine those of its activities which relate primarily to the internal affairs of science in this country, improve these activities, and extend them when and where that seems desirable in view of present circumstances. This must in particular involve a most careful review of the program and policy in respect to meetings, publications, service to scientific societies and similar groups, aids to research, and various aspects of the interrelations of science and government.

There should be explicit mention of one particularly important aspect of this internal problem of service to science. We have reached the stage where one over-all organization cannot effectively deal with the intensive and specialized interests of individual branches of science. The technical papers that present detailed results in chemistry, in physics, in mathematics, in zoology, etc., can more properly be presented before meetings sponsored and arranged by the appropriate professional groups.

It is thus clear that the AAAS should not attempt to hold to a pattern of annual meetings that was natural and effective many years ago, but which is now outmoded. This is, in fact, only one aspect of an important general principle. In view of the present size and complexity of science, in view of the seriousness and importance of the relation of science to society, and in view of the unique inclusiveness of the AAAS, it seems clear that this organization should devote less of its energies to the more detailed and more isolated technical aspects of science, and devote more of its energies to broad problems that involve the whole of science, the relations of science to government, indeed the relations of science to our society as a whole.

This increased emphasis on broad problems should lead to new activities in wider fields, but it also requires a modification of what the AAAS tries to do within and for science. Thus it seems clear that a major present opportunity for the AAAS within science is to act, in all ways that promise useful results, as a synthesizing and unifying influence. As an obvious example, this indicates meetings at which one branch of science is interpreted to the other branches of science, meetings which cultivate borderline fields, and meetings at which the unifying theme would be central problems whose treatment requires the attack of several disciplines.

This opportunity to try to "put science back together" seems so important that it may be wise to modify the existing statement (quoted in the next paragraph) of the purpose of the AAAS to include more specific dedication to synthesizing activities. Such activities are, of course, wholly consistent with the present statement of purpose; but if in fact this is, as some of us feel to be the case, the major present internal opportunity of the AAAS, then it deserves explicit statement.

Finally, this same emphasis on over-all problems demands that the AAAS not only recognize but attack the broader external problem of the relation of science to society. It seems to us necessary that the AAAS now begin to take seriously one statement of purpose that has long existed in its constitution. To quote:

The objects of the American Association for the Advancement of Science are to further the work of scientists, to facilitate cooperation among them, to improve the effectiveness of science in the promotion of human welfare, *and to increase public understanding and appreciation of the importance and promise of the methods of science in human progress.*

It is clearly recognized that the diffusion among the general public of knowledge about science and its methods is a difficult, slow, and never-ending job. It would require staff, money, patience, and wisdom. It would involve failures, and it would at some points strain the professional sensitivities of scientists. But in our modern society it is absolutely essential that science—the results of science, the nature and importance of basic research, the methods of science, the spirit of science—be better understood by government officials, by businessmen, and indeed by all the people.

We enthusiastically reaffirm our belief in the statement quoted just above as the culminating object of the AAAS; and we favor the adoption, after suitable study, of activities in this field as a major active interest of the AAAS.

It is recommended that this tentative statement of general policy be placed before the whole membership of the AAAS; and that it be sent to all the members of the Council, accompanied by a request for serious consideration and response. As soon as democratic procedures indicate that the above statement, or some then available modification of it, represents a consensus, the Executive Committee should set up a series of committees to study the practical and detailed problems of implementing these principles.

These studies themselves should be carried out in a scientific manner with disregard of vested interests, with tempered concern for traditional procedures, and with imagination in respect to the present future. Various aspects of the studies can doubtless be usefully aided by the methods of operations research, so that judgments can be at least partially founded on fact, as well as opinion.

Executive Committee of AAAS: Roger Adams, Detlev W. Bronk, John R. Dunning, E.B. Fred, Walter S. Hunter, Paul E. Klopsteg, Kirtley F. Mather, Howard A. Meyerhoff, Fernandus Payne, Paul B. Sears, Warren Weaver.

Special Consultants: Philip B. Armstrong, J. H. Bodine, A. J. Carlson, Mervin J. Kelly, John E. Pfeiffer, Ralph A. Rohweder, E. C. Stakman, Marshall H. Stone, Alan T. Waterman.

Arden House, 15 September 1951

APPENDIX B:
ADMINISTRATIVE OFFICERS, 1848–1998

Principal Administrative Officers

PERMANENT SECRETARIES

Spencer F. Baird	1851–1854
Joseph Lovering	1854–1868
F. W. Putnam	1869
Joseph Lovering	1870–1873
F. W. Putnam	1873–1898
L. O. Howard	1898–1920
Burton E. Livingston	1920–1930
Charles F. Roos	1931–1932
Henry B. Ward	1933–1937
F. R. Moulton	1937–1946

ADMINISTRATIVE SECRETARIES

F. R. Moulton	1946–1948
Howard A. Meyerhoff	1949–1953
Dael Wolfle	1954–1955

EXECUTIVE OFFICERS

Dael Wolfle	1956–1970
William Bevan	1970–1974
Philip H. Abelson (acting)	1974–1975
William D. Carey	1975–1987
Alvin W. Trivelpiece	1987–1988
Philip H. Abelson (acting)	1989
Richard S. Nicholson	1989–

ASSOCIATE ADMINISTRATIVE OFFICERS

F. S. Hazard, Assistant Secretary	1912–1920

Sam Woodley, Assistant Secretary	1920–1945
Howard A. Meyerhoff, Executive Secretary	1945–1946
J. M. Hutzel, Assistant Administrative Secretary	1946–1948
Raymond L. Taylor, Assistant Administrative Secretary	1949–1953
Associate Administrative Secretary	1953–1967
John A. Behnke, Assistant Administrative Secretary	1952–1953
Associate Administrative Secretary	1953–1956
William T. Kabisch, Assistant Executive Officer	1967–1970
Richard Trumbull, Deputy Executive Officer	1970–1974
J. Thomas Ratchford, Associate Executive Officer	1977–1989

TREASURERS

Jeffries Wyman	1848
A. L. Elwyn	1849–1870
William S. Vaux	1871–1881
William Lilly	1882–1893
R. S. Woodward	1894–1924
John L. Wirt	1925–1940
Carroll W. Morgan	1941–1944
William E. Wrather	1945–1953
Paul A. Scherer	1954–1962
Paul E. Klopsteg	1963–1969
William T. Golden	1969–

Editors of Science

Science was founded in 1880 by Thomas A. Edison. John Michels was the first editor. In 1883, Alexander Graham Bell and Gardiner Greene Hubbard purchased the magazine and established the Science Company, which published *Science* from 1883 to 1894. The magazine was then sold to James McKeen Cattell. In 1900, the AAAS entered into an agreement with Dr. Cattell to make *Science* the official journal of the Association, and, in 1945, the AAAS became its owner and publisher.

James McKeen Cattell	1895–1944
Josephine Owen Cattell and Jaques Cattell	1944–1945
Willard L. Valentine	1946–1947
Mildred Atwood (acting)	1947–1948
Editorial Board, George Baitsell, chair	1948–1949
Howard A. Meyerhoff	1949–1953
H. Bentley Glass (acting)	1953
Duane Roller	1954
Dael Wolfle (acting)	1955
Graham DuShane	1956–1962
Philip H. Abelson	1962–1984

Daniel E. Koshland, Jr.	1985–1995
Floyd E. Bloom	1995–

Former Members of the Board of Directors (excluding presidents), 1964–1996

Robert McC. Adams	1985–1988
Mary Ellen Avery	1989–1993
Francisco J. Ayala	1989–1993
Robert W. Berliner	1983–1987
David Blackwell	1970–1972
Floyd E. Bloom	1986–1990
Richard H. Bolt	1969–1976
Lewis M. Branscomb	1970–1973
Eloise E. Clark	1978–1981
Kenneth B. Clark	1975–1976
Mary E. Clutter	1986–1990
Joel E. Cohen	1976
Barry Commoner	1967–1974
Eugene H. Cota-Robles	1988–1992
Martin M. Cummings	1977–1980
Ruth M. Davis	1974–1977
Mildred S. Dresselhaus	1985–1989
Renée C. Fox	1977–1980
Robert A. Frosch	1990–1994
John W. Gardner	1964–1965
Joseph G. Gavin, Jr.	1988–1992
John H. Gibbons	1988–1991
Bernard Gifford	1977–1978
David R. Goddard	1964–1967
Nancie L. Gonzalez	1980–1984
Ward H. Goodenough	1972–1975
Beatrix A. Hamburg	1987–1991
Florence P. Haseltine	1990–1994
Caryl P. Haskins	1971–1974
Hudson Hoagland	1966–1969
Gerald Holton	1967–1970
William A. Lester, Jr.	1993–1997
Simon A. Levin	1994–1998
Mike McCormack	1976–1979
Daniel P. Moynihan	1972–1973
Dorothy Nelkin	1983–1987
Phyllis V. Parkins	1970–1973
Russell W. Peterson	1978–1981
Anna C. Roosevelt	1994–1998
John C. Sawhill	1979–1980

John E. Sawyer	1982–1986
Jean'ne M. Shreeve	1991–1995
Alan Schriesheim	1992–1996
Chauncey Starr	1974–1977
H. Burr Steinbach	1964–1969
Jean E. Taylor	1995–1999
Kenneth V. Thimann	1968–1971
Chang-Lin Tien	1992–1996
Warren Washington	1991–1995
Nancy S. Wexler	1993–1997
John A. Wheeler	1965–1968
Linda S. Wilson	1984–1988
Chen Ning Yang	1976–1979
Harriet Zuckerman	1980–1984

APPENDIX C:
AAAS MEMBERSHIP, 1848–1998

Year	Membership	Year	Membership
1848	461	1875	807
1849	542	1876	867
1850	800	1877	953
1851	800	1878	962
1852	*	1879	1,030
1853	940	1880	1,555
1854	1,004	1881	1,699
1855	605	1882	1,922
1856	722	1883	1,981
1857	946	1884	1,981
1858	962	1885	1,956
1859	862	1886	1,886
1860	644	1887	1,956
1861	*	1888	1,964
1862	*	1889	1,952
1863	*	1890	1,944
1864	*	1891	2,054
1865	*	1892	2,037
1866	637	1893	1,939
1867	415	1894	1,802
1868	686	1895	1,913
1869	511	1896	1,890
1870	536	1897	1,782
1871	668	1898	1,729
1872	610	1899	1,721
1873	670	1900	1,925
1874	722	1901	2,703

Year	Membership	Year	Membership
1902	3,474	1945	27,175
1903	4,005	1946	28,725
1904	4,041	1947	33,442
1905	4,321	1948	42,545
1906	4,498	1949	44,947
1907	5,114	1950	46,775
1908	6,136	1951	48,493
1909	7,950	1952	48,740
1910	*	1953	48,803
1911	8,041	1954	48,660
1912	8,333	1955	50,189
1913	8,350	1956	52,718
1914	8,325	1957	55,727
1915	*	1958	57,150
1916	*	1959	59,292
1917	*	1960	62,097
1918	*	1961	68,890
1919	11,442	1962	79,750
1920	11,547	1963	79,260
1921	11,547	1964	88,393
1922	11,646	1965	98,307
1923	11,707	1966	109,994
1924	12,887	1967	117,156
1925	12,263	1968	122,561
1926	14,263	1969	127,717
1927	14,862	1970	133,364
1928	16,328	1971	130,485
1929	18,462	1972	127,402
1930	19,059	1973	126,686
1931	19,889	1974	120,458
1932	18,665	1975	115,411
1933	18,549	1976	113,454
1934	18,553	1977	127,832
1935	18,102	1978	127,520
1936	18,242	1979	128,262
1937	18,776	1980	131,582
1938	19,059	1981	135,139
1939	20,195	1982	137,453
1940	21,067	1983	134,400
1941	21,798	1984	135,961
1942	23,671	1985	131,238
1943	24,726	1986	131,260
1944	25,487	1987	133,114

Year	Membership	Year	Membership
1988	132,366	1994	143,230
1989	128,187**	1995	144,005
1990	133,033	1996	146,604
1991	131,170	1997	146,212
1992	136,864	1998	141,201
1993	141,840		

* No data. AAAS did not meet this year
** Drop due to delayed renewals
Source: AAAS Office of Membership and Meetings

APPENDIX D:
AAAS MEETINGS AND PRESIDENTS

1848 Sept.	Philadelphia	William B. Rogers (geology) [acting until the installation of first elected president, Redfield] William C. Redfield (geology)
1849 Aug.	Cambridge	Joseph Henry (physics)
1850 Mar.	Charleston	A. D. Bache (oceanography)
1850 Aug.	New Haven	A. D. Bache
1851 May	Cincinnati	Louis Agassiz (glaciology, zoology)
1851 Aug.	Albany	Louis Agassiz
1852	no meeting	Benjamin Peirce (mathematics)
1853 July	Cleveland	Benjamin Peirce
1854 April	Washington	James D. Dana (geology)
1855 Aug.	Providence	John Torrey (botany)
1856 Aug.	Albany	James Hall (geology)
1857 Aug.	Montreal	J. W. Bailey (chemistry) Alexis Caswell (astronomy) [successor to J. W. Bailey, deceased]
1858 April	Baltimore	Jeffries Wyman (medicine)
1859 Aug.	Springfield, Mass.	Stephen Alexander (astronomy)
1860 Aug.	Newport	Isaac Lea (geology)

(no meetings and no presidents 1861–1865)

1866 Aug.	Buffalo	F.A.P. Barnard (astronomy)
1867 Aug.	Burlington	J. S. Newberry (geology)
1868 Aug.	Chicago	Benjamin A. Gould (astronomy)
1869 Aug.	Salem, Mass.	J. W. Foster (geography)
1870 Aug.	Troy	William Chauvenet (astronomy) T. Sterry Hunt (geology) [successor to W. Chauvenet, deceased]
1871 Aug.	Indianapolis	Asa Gray (botany)

1872 Aug.	Dubuque	J. Lawrence Smith (chemistry)
1873 Aug.	Portland, ME	Joseph Lovering (physics)
1874 Aug.	Hartford	John L. LeConte (entomology)
1875 Aug.	Detroit	Julius L. Hilgard (geography)
1876 Aug.	Buffalo	William B. Rogers (geology)
1877 Aug.	Nashville	Simon Newcomb (astronomy)
1878 Aug.	St. Louis	O. C. Marsh (paleontology)
1879 Aug.	Saratoga Springs	George F. Barker (chemistry)
1880 Aug.	Boston	Lewis H. Morgan (anthropology)
1881 Aug.	Cincinnati	George J. Brush (geology)
1882 Aug.	Montreal	J. W. Dawson (geology)
1883 Aug.	Minneapolis	Charles A. Young (astronomy)
1884 Sept.	Philadelphia	J. P. Lesley (geology)
1885 Aug.	Ann Arbor	H. A. Newton (mathematics)
1886 Aug.	Buffalo	Edward S. Morse (zoology)
1887 Aug.	New York	S. P. Langley (physics)
1888 Aug.	Cleveland	J. W. Powell (geology)
1889 Aug.	Toronto	T. C. Mendenhall (physics)
1890 Aug.	Indianapolis	George L. Goodale (botany)
1891 Aug.	Washington	Albert B. Prescott (chemistry)
1892 Aug.	Rochester	Joseph LeConte (geology)
1893 Aug.	Madison	William Harkness (astronomy)
1894 Aug.	Brooklyn	Daniel G. Brinton (anthropology)
1895 Aug.	Springfield, Mass.	Edward W. Morley (chemistry)
1896 Aug.	Buffalo	Edward D. Cope (paleontology)
		Theodore Gill (zoology)
		[successor to Edward D. Cope, deceased]
1897 Aug.	Detroit	Wolcott Gibbs (chemistry)
		[WJ McGee presided in Gibbs's absence]
1898 Aug.	Boston	F. W. Putnam (anthropology)
1899 Aug.	Columbus	Edward Orton (geology)
		Marcus Benjamin (social sciences) and
		Grove Karl Gilbert (geology) [successors to
		Edward Orton, deceased]
1900 June	New York	R. S. Woodward (mathematics)
1901 Aug.	Denver	Charles S. Minot (medicine)
1902 June	Pittsburgh	Asaph Hall (astronomy)
1902 Dec.	Washington	Ira Remsen (chemistry)
1903 Dec.	St. Louis	Carroll D. Wright (economics)
1904 Dec.	Philadelphia	W. G. Farlow (botany)
1905 Dec.	New Orleans	C. M. Woodward (mathematics)
1906 June	Ithaca	William H. Welch (medicine)
1906 Dec.	New York	William H. Welch
1907 Dec.	Chicago	E. L. Nichols (physics)
1908 June	Hanover	Thomas C. Chamberlin (geology)
1908 Dec.	Baltimore	Thomas C. Chamberlin

1909 Dec.	Boston	David Starr Jordan (biology)
1910 Dec.	Minneapolis	A. A. Michelson (physics)
1911 Dec.	Washington	Charles E. Bessey (botany)
1912 Dec.	Cleveland	E. C. Pickering (astronomy)
1913 Dec.	Atlanta	Edmund B. Wilson (zoology)
1914 Dec.	Philadelphia	Charles W. Eliot (education)
1915 Aug.	San Francisco	W. W. Campbell (astronomy)
1915 Dec.	Columbus	W. W. Campbell
1916 Dec.	New York	Charles R. Van Hise (geology)
1917 Dec.	Pittsburgh	Theodore W. Richards (chemistry)
1918 Dec.	Baltimore	John Merle Coulter (botany)
1919 Dec.	St. Louis	Simon Flexner (medicine)
1920 Dec.	Chicago	Leland O. Howard (entomology)
1921 Dec.	Toronto	Eliakim H. Moore (mathematics)
1922 June	Salt Lake City	J. Playfair McMurrich (anatomy)
1922 Dec.	Boston	J. Playfair McMurrich
1923 Sept.	Los Angeles	Charles D. Walcott (paleontology)
1923 Dec.	Cincinnati	Charles D. Walcott
1924 Dec.	Washington	J. McKeen Cattell (psychology)
1925 June	Boulder	Michael I. Pupin (engineering)
1925 June	Portland, Oreg.	Michael I. Pupin
1925 Dec.	Kansas City	Michael I. Pupin
1926 Dec.	Philadelphia	Liberty Hyde Bailey (horticulture)
1927 Dec.	Nashville	Arthur A. Noyes (chemistry)
1928 Dec.	New York	Henry F. Osborn (paleontology)
1929 Dec.	Des Moines	Robert A. Millikan (physics)
1930 Dec.	Cleveland	Thomas H. Morgan (genetics)
1931 June	Pasadena	Franz Boas (anthropology)
1931 Dec.	New Orleans	Franz Boas
1932 June	Syracuse	John Jacob Abel (pharmacology)
1932 Dec.	Atlantic City	John Jacob Abel
1933 June	Chicago	Henry N. Russell (astronomy)
1933 Dec.	Boston	Henry N. Russell
1934 June	Berkeley	Edward L. Thorndike (psychology)
1934 Dec.	Pittsburgh	Edward L. Thorndike
1935 June	Minneapolis	Karl T. Compton (physics)
1935 Dec.	St. Louis	Karl T. Compton
1936 June	Rochester	Edwin G. Conklin (biology)
1936 Dec.	Atlantic City	Edwin G. Conklin
1937 June	Denver	George D. Birkhoff (mathematics)
1937 Dec.	Indianapolis	George D. Birkhoff
1938 June	Ottawa	Wesley C. Mitchell (economics)
1938 Dec.	Richmond	Wesley C. Mitchell
1939 June	Milwaukee	Walter B. Cannon (physiology)
1939 Dec.	Columbus	Walter B. Cannon
1940 June	Seattle	Albert F. Blakeslee (genetics)

1940 Dec.	Philadelphia	Albert F. Blakeslee
1941 June	Durham	Irving Langmuir (chemistry)
1941 Sept.	Chicago	Irving Langmuir
1941 Dec.	Dallas	Irving Langmuir
1942	no meeting	Arthur H. Compton (physics)
1943	no meeting	Isaiah Bowman (geography)
1944 Sept.	Cleveland	Anton J. Carlson (physiology)
1946 Mar.*	St. Louis	C. F. Kettering (engineering)
1946 Dec.	Boston	James B. Conant (chemistry)
1947 Dec.	Chicago	Harlow Shapley (astronomy)
1948 Sept.	Washington	Edmund W. Sinnott (botany)
1949 Dec.	New York	E. C. Stakman (plant pathology)
1950 Dec.	Cleveland	Roger Adams (chemistry)
1951 Dec.	Philadelphia	Kirtley F. Mather (geology)
1952 Dec.	St. Louis	Detlev W. Bronk (physiology)
1953 Dec.	Boston	Edward U. Condon (physics)
1954 Dec.	Berkeley	Warren Weaver (mathematics)
1955 Dec.	Atlanta	George W. Beadle (genetics)
1956 Dec.	New York	Paul B. Sears (plant ecology)
1957 Dec.	Indianapolis	Laurence H. Snyder (genetics)
1958 Dec.	Washington	Wallace R. Brode (chemistry)
1959 Dec.	Chicago	Paul E. Klopsteg (physics)
1960 Dec.	New York	Chauncey D. Leake (pharmacology, history of science)
1961 Dec.	Denver	Thomas Park (animal ecology)
1962 Dec.	Philadelphia	Paul M. Gross (chemistry)
1963 Dec.	Cleveland	Alan T. Waterman (physics)
1964 Dec.	Montreal	Laurence M. Gould (geology)
1965 Dec.	Berkeley	Henry Eyring (chemistry)
1966 Dec.	Washington	Alfred S. Romer (paleontology)
1967 Dec.	New York	Don K. Price (political science)
1968 Dec.	Dallas	Walter Orr Roberts (astronomy, meteorology)
1969 Dec.	Boston	H. Bentley Glass (genetics)
1970 Dec.	Chicago	Athelstan Spilhaus (meteorology, oceanography)
1971 Dec.	Philadelphia	Mina Rees (mathematics)
1972 Dec.	Washington	Glenn T. Seaborg (chemistry)
1973 June	Mexico City (special meeting)	Leonard M. Rieser (physics)
1974 Feb.	San Francisco	Roger Revelle (oceanography, geophysics, human population)
1975 Jan.	New York	Margaret Mead (anthropology)
1976 Feb.	Boston	William D. McElroy (biology, biochemistry)
1977 Feb.	Denver	Emilio Q. Daddario (law, public service)
1978 Feb.	Washington	Emilio Q. Daddario **

1979 Jan.	Houston	Edward E. David, Jr. (physics, electrical engineering)
1980 Jan.	San Francisco	Kenneth E. Boulding (economics)
1981 Jan.	Toronto	Frederick Mosteller (statistics)
1982 Jan.	Washington	D. Allan Bromley (physics)
1983 May	Detroit	E. Margaret Burbidge (astronomy)
1984 May	New York	Anna J. Harrison (chemistry)
1985 May	Los Angeles	David A. Hamburg (health science policy)
1986 May	Philadelphia	Gerard Piel (science journalism)
1987 Feb.	Chicago	Lawrence Bogorad (plant biology)
1988 Feb.	Boston	Sheila E. Widnall (aeronautics and astronautics)
1989 Jan.	San Francisco	Walter E. Massey (physics)
1990 Feb.	New Orleans	Richard C. Atkinson (psychology, cognitive science)
1991 Feb.	Washington	Donald N. Langenberg (physics)
1992 Feb.	Chicago	Leon M. Lederman (physics)
1993 Feb.	Boston	F. Sherwood Rowland (chemistry)
1994 Feb.	San Francisco	Eloise E. Clark (biology)
1995 Feb.	Atlanta	Francisco J. Ayala (evolutionary biology)
1996 Feb.	Baltimore	Rita R. Colwell (biology)
1997 Feb.	Seattle	Jane Lubchenco (ecology)
1998 Feb.	Philadelphia	Mildred Dresselhaus (engineering)

* 1945 meeting postponed
** Served as president from 1 January 1977 through 17 February 1978. Two meetings were held during his presidency.

NOTES

Creating a Forum for Science: AAAS in the Nineteenth Century

1. The most thorough accounts of this period and organizational developments are found in Sally Gregory Kohlstedt, *The Formation of the American Scientific Community: The American Association for the Advancement of Science, 1848–1860* (Urbana: University of Illinois Press, 1976); Robert V. Bruce, *The Launching of Modern American Science, 1846–1876* (New York: Knopf, 1987); and A. Hunter Dupree, *Science in the Federal Government: A History of Policies and Activities to 1940* (Baltimore: Johns Hopkins University Press, 1987 [1st ed., 1957].)
2. Sally Gregory Kohlstedt, "A Step toward Scientific Self-Identity in the United States: The Failure of the National Institute," *Isis* 62 (fall 1971): 339–362.
3. Jane Colden, *The Botanic Manuscript of Jane Colden*, ed. H. W. Rickett and Elizabeth C. Hall (New York: Garden Club of Orange and Duchess Counties, 1963); and Silvio A. Bedini, *The Life of Benjamin Banneker* (New York: Scribner's, 1972).
4. Alexandra Oleson and Sanborn C. Brown, eds., *The Pursuit of Knowledge in the Early American Republic: American Scientific and Learned Societies from Colonial Times to the Civil War* (Baltimore: Johns Hopkins University Press, 1976).
5. George P. Merrill, *Contributions to a History of American State Geological and Natural History Surveys*, Smithsonian Institution, Bulletin 109 (Washington, D.C., 1920); and Patsy Gerstner, *Henry Darwin Rogers, 1808–1866: American Geologist* (Tuscaloosa: University of Alabama Press, 1994).
6. George H. Daniels, *American Science in the Age of Jackson* (New York: Columbia University Press,1968); and Sally Gregory Kohlstedt, "Savants and Professionals: The American Association for the Advancement of Science, 1848–1860," in *The Pursuit of Knowledge in the Early American Republic: American Scientific and Learned Societies from Colonial Times to the Civil War*, ed. Alexandra Oleson and Sanborn C. Brown (Baltimore: Johns Hopkins University Press, 1976), pp. 299–325.
7. Sally Gregory Kohlstedt, "The Geologists' Model for National Science, 1840–

1847," *Proceedings of the American Philosophical Society* 118 (April 1974): 179–195.

8. Association of American Geologists and Naturalists, *Report* (1843): 70–71, 474–515.

9. William Barton Rogers to J. W. Bailey, 22 October 1843, Rogers Papers, Massachusetts Institute of Technology Archives.

10. A copy of the printed "Circular" dated 10 May 1848 is in the AAAS Archives, Washington, D.C.

11. Jack Morrell and Arnold Thackray, *Gentlemen of Science: Early Years [1831–1844] of the British Association for the Advancement of Science* (Oxford: Clarendon Press, 1981); and Roy Macleod and Peter Collins, eds., *The Parliament of Science: The British Association for the Advancement of Science, 1831–1981* (Northwood, U.K.: Science Reviews Ltd., 1981).

12. Henry Rogers to William Rogers, 16 May 1848, in Emma Rogers, ed., *Life and Letters of William Barton Rogers* (Boston: Houghton, Mifflin, and Co., 1896), I: 287–288.

13. Lillian B. Miller, *The Lazzaroni: Science and Scientists in Mid-Nineteenth Century America* (Washington: National Portrait Gallery, 1972); and Emma Rogers, ed., *Life and Letters of William Barton Rogers*, 2 vols. (Boston, Mass.: Houghton, Mifflin, and Co., 1896).

14. Hugh Richard Slotten, *Patronage, Practice, and the Culture of American Science: Alexander Dallas Bache and the U.S. Coast Survey* (Cambridge, U.K.: Cambridge University Press, 1994).

15. Proceedings of the AAAS, 4 (1851): lii–liii.

16. The membership list is not an index to practicing scientists, since some members obviously joined because of an interest rather than involvement. Estimates of the size of the scientific community in this period vary, with Robert Bruce identifying the 1,078 scientists and technologists listed in the *Dictionary of American History* between 1846 and 1876 as one cohort. See his *Launching of Modern American Science*, p. 363.

17. Lester D. Stephens, *Joseph LeConte: Gentle Prophet of Evolution* (Baton Rouge: Louisiana State University Press, 1982), pp. 72–73.

18. Fillmore to Joseph Henry, 4 August 1866, Indiana Historical Society, Indianapolis, Indiana.

19. Quoted in Helen Wright, *Sweeper in the Sky: The Life of Maria Mitchell, First Woman Astronomer in America* (New York, Macmillan, 1949), p. 69.

20. *New York Times*, 3 May 1858.

21. *Science* 6, no. 136 (1885): 201.

22. Edward Lurie, *Louis Agassiz: A Life in Science* (Chicago: University of Chicago Press, 1960), pp. 260–261; and William R. Stanton, *The Leopard's Spots: Scientific Attitudes toward Race in America, 1815–1859* (Chicago: University of Chicago Press, 1960).

23. Quoted in Bruce, *Launching of Modern American Science*, p. 258.

24. Lester D. Stephens, *Ancient Animals and Other Wondrous Things: The Story of Francis Simmons Holmes, Paleontologist and Curator of the Charleston Mu-*

seum (Charleston: Contributions from the Charleston Museum, no. 17 (1988), pp. 13–17.

25. Tamara Haygood, *Henry William Ravenel, 1814–1887: South Carolina Scientist in the Civil War Era* (Tuscaloosa: University of Alabama press, 1987), pp. 58–61.

26. In honor of this occasion Alfred B. Street penned *Science: A Poem* (Albany: Van Benthuysen, 1856) dedicated to the AAAS; Michele Aldrich kindly brought this pamphlet to my attention.

27. Henry D. Shapiro, "The Western Academy of Natural Sciences of Cincinnati and the Structure of Science in the Ohio Valley, 1810–1850," in *The Pursuit of Knowledge in the Early American Republic: American Scientific and Learned Societies from Colonial Times to the Civil War*, ed. Alexandra Oleson and Sanborn C. Brown (Baltimore: Johns Hopkins University Press, 1976), pp. 219–247.

28. *Proceedings of the AAAS* 17 (1868): 355–359. For clarity, the date in parentheses will indicate the year of the meeting, not the year of publication, which most often was in the year following the annual meeting.

29. Rexmond C. Cochrane, *The National Academy of Sciences: The First Hundred Years, 1863–1963* (Washington, D.C.: National Academy of Sciences, 1978).

30. Nathan Reingold, *Science, American Style* (New Brunswick: Rutgers University Press, 1991), pp. 156–168.

31. The address is now available in Arthur P. Molella et al., eds., *A Scientist in American Life: Essays and Lectures of Joseph Henry* (Washington, D.C.: Smithsonian Institution Press, 1980), pp. 35–50.

32. John D. Holmfeld, "From Amateurs to Professionals in American Science: the Controversy over the Proceedings of an 1853 Scientific Meeting," *Proceedings of the American Philosophical Society* 114 (1970): 22–36.

33. Nathan Reingold, "Definitions and Speculations: The Professionalization of Science in America in the Nineteenth Century," in *The Pursuit of Knowledge*, pp. 33–69 .

34. Thoreau's response to the scientists is recorded in Henry D. Thoreau, *Journal*, ed. Patrick F. O'Connell (Princeton: Princeton University Press, 1897), pp. 469–470.

35. Margaret Rossiter, *Women Scientists in America: Struggles and Strategies to 1940* (Baltimore: Johns Hopkins University Press, 1982), pp. 76–78.

36. Harry G. Lang, *Science of the Spheres: The Deaf Experience in the History of Science* (Westport: Bergin and Garvey, 1994), esp. chapter 2.

37. [Rogers to Hall], 30 January 1866, Rogers Papers, Massachusetts Institute of Technology Archives.

38. Ronald L. Numbers and Todd L. Savitt, eds., *Science and Medicine in the Old South* (Baton Rouge: Louisiana State University Press, 1989).

39. Daniel J. Kevles, *The Physicists: The History of a Scientific Community in Modern America* (Cambridge: Harvard University Press, 1995 [1st ed., 1978]), p. 41.

40. Stephens, *Joseph LeConte*, pp. 206–208.

41. Rossiter, *Women Scientists in America*, pp. 76–78.

42. Erminnie Smith had already been the first woman secretary, also in the Anthropology Section. See *Proceedings of the AAAS* 33 (1884): 707.

43. John Lankford, *American Astronomy: Community, Careers, and Power, 1859–1940* (Chicago: University of Chicago Press, 1997), pp. 23–26.

44. A. Hunter Dupree, *Asa Gray, 1810–1888* (Cambridge: Harvard University Press, 1959).

45. *Proceedings of the AAAS* 21 (1872): 2.

46. "On the Origin of Species," *Proceedings of the AAAS* 22 (1873): 7–12.

47. "Change by Gradual Modification to the Universal Law," *Proceedings of the AAAS* 36 (1887): 73–76.

48. *Proceedings of the AAAS* (1976): 138. Morse's address was abridged and reprinted in *Science* (12 August 1887).

49. Albert E. Moyer, *A Scientist's Voice in American Culture: Simon Newcomb and the Rhetoric of Scientific Method* (Berkeley: University of California Press, 1992), esp. chapter 8.

50. For a useful discussion of the idea, see Nathan Reingold, "American Indifference to Basic Research: A Reappraisal," In George Daniels, ed., *Nineteenth-Century American Science; A Reappraisal* (Evanston, Ill.: Northwestern University Press, 1972), pp. 38–62.

51. Kevles, *The Physicists*, 43.

52. *Proceedings of the AAAS* 36 (1887): 120.

53. Dupree, *Science in the Federal Government*, pp. 115–116.

54. Kohlstedt, *Formation*, p. 97.

55. Quoted in Kohlstedt, *Formation*, p. 122.

56. Wright, *Sweeper in the Sky*, p. 69.

57. *Proceedings of the AAAS* 36 (1887): 349; 37 (1888): 38–41; 39 (1890): 21 43 (1894): 464.

58. John F. Reiger, *American Sportsmen and the Origins of Conservation*, rev. ed. (Norman: University of Oklahoma Press, 1986 [1st ed., 1975]).

59. *Proceedings of the AAAS* 37 (1888): 359–367; 38 (1889): 447; 39 (1890): 432–433; 40 (1891): 446–447; 41 (1892): 353–354. The capstone to Fernow's efforts was his vice presidential address on "The Providential Functions of Government with Special Reference to Natural Resources," *Proceedings of the AAAS* 44 (1895): 325–344.

60. Kohlstedt, *Formation*, pp. 130–131.

61. For various discussions of these matters see *Proceedings of the AAAS* 15 (1866): 120; 45 (1896): 294; 24 (1875): 19–24.

62. *Science* magazine often had criticisms and comparisons to make, as in vol. 6, no. 136 (1885): 202; and vol. 8, no. 187 (1886): 200.

63. Paul Farber, "Thomas Huxley and Educational Literacy," presentation at the AAAS meeting in Philadelphia, 15 February 1998.

64. For a thorough and contextual discussion of these events, see W. Conner Sorensen, *Brethren of the Net: American Entomology, 1840–1880* (Tuscaloosa: University of Alabama Press, 1995).

65. Sally Gregory Kohlstedt, "*Science*: The Struggle for Survival, 1880–1894," in

The Science Centennial Review, ed. Philip H. Abelson and Ruth Kulstad (Washington, D.C.: AAAS, 1980), pp. 15–24.

66. *Proceedings of the AAAS* 29 (1880): 738.
67. *Proceedings of the AAAS* 30 (1881): lix.
68. Quoted in Bruce, *Launching of Modern American Science*, p. 356.
69. "American Association for the Advancement of Science," *Science* 4 (1884): 229.
70. Quoted in Kohlstedt, "*Science*: The Struggle for Survival," p. 18.
71. Howard S. Miller, *Dollars for Research: Science and Its Patrons in Nineteenth-Century America* (Seattle: University of Washington Press, 1970).
72. *Proceedings of the AAAS* 23 (1874): 422.
73. Quoted in Miller, *Dollars for Research*, p. 127.
74. "A Comparative Study of the Associations," *Science* 4, no. 84 (1894): 2271–2272.
75. *Proceedings of the AAAS*, 46 (1897): 82–83.
76. David H. Guston, "Congressmen and Scientists in the Making of Science Policy: The Allison Commission, 1884–1886," *Minerva* 32 (1994): 25–52.
77. *Proceedings of the AAAS* 34 (1885): 546. Also see Stephen Skowronek, *Building a New American State: The Expansion of National Administrative Capacities, 1877–1920* (New York: Cambridge University Press, 1982); and Margaret Susan Thompson, *The "Spider Web:" Congress and Lobbying in the Age of Grant* (Ithaca: Cornell University Press, 1985). Political historians suggest that the retrenchment under the Cleveland administration also played some role, and the new civil service policies changed older patterns of recruitment and participation in the scientific agencies.
78. Alfred Tozzer, "Frederic Ward Putnam," *Biographical Memoirs of the National Academy of Sciences* 16 (1936): 125–153.
79. Ralph W. Dexter, "F. W. Putnam as Secretary of the American Association for the Advancement of Science (1873–1898)," *Essex Institute Historical Collections* 118 (1982): 106–118.
80. For a rare view of finances and management issues see Putnam to William Barton Rogers, 18 June 1877, Massachusetts Institute of Technology Archives. On Watson's retirement see *Proceedings of the AAAS* 44 (1895): 394.
81. Quoted in Dexter, "Putnam as Secretary," p. 109.
82. Nancy Smith Midgette, *To Foster the Spirit of Professionalism: Southern scientists and State Academies of Science* (Tuscaloosa: University of Alabama Press, 1991).
83. Simon Baatz, *Knowledge, Culture, and Science in the Metropolis: The New York Academy of Sciences, 1817–1970*, Annals of the New York Academy of Sciences, 584 (1990).
84. Discussion of the issues relating to the configuration of sections are found in the reports of the general secretary for every meeting as well as in Frederic Ward Putnam's annual report, both published at the end of each volume of *Proceedings*. The constitution, as amended during the year's annual meeting, is printed in the prefatory materials in each volume. For quotations and details of the changes presented here, see *Proceedings of the AAAS* 29(1880): 738, 746 and

757; 30 (1881): 372; 31 (1882): xxi–xxii; 34 (1885): 545; 36 (1887): 345; and 42 (1893): xxxiv.

85. Elizabeth B. Keeney, *The Botanizers: Amateur Scientists in Nineteenth-Century America* (Chapel Hill: University of North Carolina Press, 1992).

86. Toby Appel, "Organizing Biology: The American Society of Naturalists and its 'Affiliated Societies,' 1883–1923," in *The American Development of Biology*, ed. Ronald Rainger, Keith R. Benson, and Jane Maienschein (Philadelphia: University of Pennsylvania Press, 1988), pp. 87–120.

87. *Proceedings of the AAAS* 37 (1888): 418.

88. *Proceedings of the AAAS* 39 (1890): 3.

89. Edward Orton, "Proper Objects of the American Association for the Advancement of Science," *Popular Science Monthly* 55 (August 1899): 466–472.

90. Quoted in Curtis M. Hinsley, Jr., *Savages and Scientists: The Smithsonian Institution and the Development of American Anthropology, 1846–1910* (Washington, D.C.: Smithsonian Institution Press, 1981), p. 234.

91. Michael M. Sokal, "Science and James McKeen Cattell, 1894–1945," in *The Science Centennial Review* (Washington, D.C.: AAAS, 1980), pp. 25–33.

92. Frederic Ward Putnam, "The History, Aims, and Importance of the American Association for the Advancement of Science," *Science*, n.s. 2, no. 33 (1895): 171–174.

93. Quoted in Bruce, *Launching of Modern American Science*, p. 362.

94. Proceedings of the AAAS 47 (1898): 624. Also see John E. Graf and Dorothy W. Graf, "Leland Ossian Howard, June 11, 1857–May 1, 1950," *Biographical Memoirs of the National Academy of Sciences* 33 (1959): 87–124.

Promoting Science in a New Century: The Middle Years of the AAAS

1. W J McGee, "Fifty Years of American Science," *The Atlantic* 82 (1898): 307–320.

2. N.D.C. Hodges, "'Science' The Journal a Gift to the Scientific World," 30 April 1894, Samuel P. Langley secretarial papers, Smithsonian Institution Archives, Washington, D.C.

3. Sally Gregory Kohlstedt, "Predecessors of *Science*," *Science* 209 (4 July 1980): 34; Kohlstedt, "*Science*: The Struggle for Survival, 1880 to 1894," *Science* 209 (4 July 1980): 33–42.

4. W J McGee et al., "To the Council of the American Association for the Advancement of Science," 22 August 1894, James McKeen Cattell papers, Manuscript Division, Library of Congress, Washington, D.C.

5. Michael M. Sokal, "*Science* and James McKeen Cattell, 1894 to 1945," *Science* 209 (1980): 43–52.

6. Robert S. Woodward to Cattell, 13 January 1920, Cattell papers.

7. Michael M. Sokal, "Life-Span Developmental Psychology and the History of Science," in *Beyond History of Science: Essays in Honor of Robert E. Schofield*, ed. Elizabeth W. Garber (Lehigh University Press, 1990), pp. 67–80.

8. Michael M. Sokal, "Baldwin, Cattell, and the *Psychological Review*: A Collaboration and Its Discontents," *History of the Human Sciences* 10 (1997): 57–89.

9. Cattell to N.D.C. Hodges, undated (late 1893), 6 June [1894], undated (late June 1894), undated (August 1894), Cattell to Gardiner G. Hubbard, undated (late June 1894), undated (August 1894), 1890s letterbooks, Cattell papers.

10. Cattell to Alexander Graham Bell, 6 November 1894, Bell papers, Manuscript Division, Library of Congress; Cattell to Edward C. Pickering, 6 November 1894, Harvard College Observatory records, Harvard University Archives, Cambridge.

11. Sokal, "*Science* and James McKeen Cattell."

12. "Editor's Table," *American Naturalist* 31 (1897): 800–801.

13. Cattell to Daniel G. Brinton, 3 November 1894, Brinton papers, University of Pennsylvania Library, Philadelphia.

14. Cattell to W. W. Campbell, 22 August 1896, Lick Observatory archives, University of California, Santa Cruz.

15. Cattell to Christine Ladd Franklin, 21 March 1896, Ladd Franklin papers, Rare Book and Manuscript Library, Butler Library, Columbia University, New York.

16. [Cattell,] "The Buffalo Meeting," *Science* 4 (4 September 1896): 277–279.

17. Charles R. Barnes, "Report of the General Secretary," *Proceedings of the AAAS* 45 (1896): 237–258; Asaph Hall, Jr., "Report of the General Secretary," *Proceedings of the AAAS* 46 (1897): 465–479.

18. Cattell to L. O. Howard, 23 September 1899, Cattell papers.

19. Charles Baskerville, "Report of the General Secretary," *Proceedings of the AAAS* 49 (1900): 371–387; Howard, "Report of the Permanent Secretary," *Proceedings of the AAAS*, pp. 387–389.

20. "The Editorial Committee of *Science*," *Science* 19 (8 January 1904): 77.

21. Frank Luther Mott, *A History of American Magazines, 1850–1865* (vol. 2) (Cambridge: Harvard University Press, 1938), pp. 316–324.

22. L. O. Howard, *Fighting the Insects* (New York: Macmillan, 1933); John E. Graf and Dorothy W. Graf, "Leland Ossian Howard," *Biographical Memoirs of the National Academy of Sciences* 33 (1959): 87–124.

23. For example, "The Winter Meetings of the Scientific Societies," *Science* 6 (31 December 1897): 988–989; "The American Association for the Advancement of Science," *Science* 12 (6 July 1900): 1–4.

24. Charles S. Minot to Robert S. Woodward, 13 March 1901, Cattell papers.

25. Daniel T. MacDougal, "Report of the General Secretary," *Proceedings of the AAAS* 51 (1902): 547–569; Henry B. Ward, "Report of the General Secretary," *Proceedings of the AAAS* 52 (1903): 539–560; "American Association for the Advancement of Science," *Nature* 67 (29 January 1903): 298–299.

26. Albert E. Moyer, "American Physics in 1887," in *The Michelson Era in American Science, 1870–1930*, ed. Stanley Goldberg and Roger H. Steuwer (New York: American Institute of Physics, 1988), pp. 102–110; Dorothy Livingston, *The Master of Light: A Biography of Albert A. Michelson* (Chicago: University of Chicago Press, 1973), pp. 139–140, 181.

27. Ronald C. Tobey, *The American Ideology of National Science, 1919–1930* (Pittsburgh: University of Pittsburgh Press, 1971), p. 10; Stanley Goldberg, *Understanding Relativity: Origin and Impact of a Scientific Revolution* (Boston: Birkhauser, 1984), pp. 241–266; Gerald Holton, "On the Hesitant Rise of Quan-

tum Physics Research in the United States," in Goldberg and Steuwer, eds., *The Michelson Era in American Science, 1870–1930*, pp. 177–205.

28. Andrew Denny Rodgers III, *John Merle Coulter: Missionary in Science* (Princeton: Princeton University Press, 1944), pp. 134, 168, 213, 216, 221, 224.

29. Henry F. Nachtrieb, "The American Association for the Advancement of Science: Section F—Is It Worth While?," *Science* 37 (7 February 1913): 199–205; Toby A. Appel, "Organizing Biology: The American Society of Naturalists and Its 'Affiliated Societies,' 1883–1923, in *The American Development of Biology*, ed. Ronald Rainger, Keith R. Benson, and Jane Maienschein (Philadelphia: University of Pennsylvania Press, 1988), pp. 87–120; Benson and C. Edward Quinn, *The American Society of Zoologists, 1889–1989: A Century of Integrating the Biological Sciences* (n.p.; American Society of Zoologists, 1989).

30. John C. Burnham, *How Superstition Won and Science Lost: Popularizing Science and Health in the United States* (New Brunswick, N.J.: Rutgers University Press, 1987), pp. 152, 308.

31. [Cattell,] "American Association for the Advancement of Science," *Science* 6 (6 August 1897): 181–184.

32. Sokal, "*Science* and James McKeen Cattell.

33. A. Hunter Dupree, *Science in the Federal Government: A History of Policies and Activities to 1940* (Cambridge: Harvard University Press, 1957), pp. 269–270.

34. Michele L. Aldrich and Alan E. Leviton, "The Dark Ages of AAAS, 1865–1900," presented at the AAAS Annual Meeting, Philadelphia, 13 February 1998; *Proceedings of the AAAS 59* (1908): 113; Margaret W. Rossiter, *Women Scientists in America: Struggles and Strategies to 1940* (Baltimore: Johns Hopkins University Press, 1982), pp. 76, 79–80, 275, 285; Dael Wolfle, *Renewing a Scientific Society: The American Association for the Advancement of Science from World War II to 1970* (Washington, D.C.: AAAS, 1989), pp. 140–142.

35. L. O. Howard, "The American Association for the Advancement of Science (1838–)," *Journal of Proceedings and Addresses of the National Educational Association*, 1906, pp. 475–479; William H. Hale, "American Association for the Advancement of Science," *Scientific American* 96 (12 January 1907): 46; [Slosson,] "Discussion in the British Association," *The Independent* 63 (5 September 1907): 579–580.

36. L. O. Howard, *Fighting the Insects: The Story of an Entomologist* (New York: Macmillan, 1933), pp. 283–284; William H. Welch to Cattell, 5 January [1907], Cattell papers.

37. "The Smithsonian Institution and the National Academy of Sciences and the American Association for the Advancement of Science," *Science* 25 (3 May 1907): 716–717; Cattell to Daniel T. MacDougal, 17 August 1907, MacDougal papers, Arizona Heritage Center, Tucson; Cattell to Minot 29 August 1907, Cattell papers; Cattell to Herman L. Fairchild, 13 November 1907, Fairchild papers, University of Rochester Library, Rochester, N.Y.

38. Minutes of the [AAAS] Committee on Policy, 29 December 1907, AAAS archives; Cattell to Howard, 23 October 1908, Fairchild to Cattell, 26 January 1909, Minot to Cattell, 26 January 1909, 9 December 1909, Howard to Cattell, 3 Sep-

tember 1909, Cattell papers; Howard, "The Annual Dues of Members of the American Association for the Advancement of Science," *Science* 28 (11 December 1908): 834; H. Newell Wardle, "Railroad Rates for the Baltimore Meeting," *Science* 29 (22 January 1909): 144.

39. Minutes of the [AAAS] Council, 30 December 1907, 28 December 1909, 30 December 1909, 27 November 1911, 2 January 1914, 21 April 1914, Minutes of the [AAAS] Committee on policy, 31 December 1907, 17 November 1909, 29 December 1909, 17 April 1911, 20 November 1911, 26 December 1911, 20 April 1914, 19 April 1915, AAAS archives; Minot to Cattell, 3 January 1911, JMC papers; Cattell to Minot, 4 January 1911, Minot "to the Members of the Committee on Policy," 29 January 1911, Fairchild papers; "The American Association for the Advancement of Science," *Science* 33 (28 April 1911): 652.

40. Minutes of the [AAAS] Committee on Policy, 11 November 1912, 29 December 1912, 3 January 1913, 21 April 1913, 17 November 1913, 20 April 1914, Minutes of the [AAAS] Council, 30 December 1912, 2 January 1913, 3 January 1913, 21 April 1913, 22 April 1913, 30 December 1913, AAAS archives; H. E. Summers, "The Cleveland Meeting of the American Association for the Advancement of Science," *Science*, 37 (10 January 1913): 41–46; "The Atlanta Meeting of the American Association for the Advancement of Science," *Science*, 38 (5 December 1913): 808–811; H. W. Springsteen, "The American Association for the Advancement of Science: The Atlanta Meeting," *Science* 39 (9 January 1914): 39–42.

41. Minutes of the [AAAS] committee on policy, 21 April 1908, 18 November 1908, 17 November 1909, 11 November 1912, Minutes of the [AAAS] Council, 18 April 1911, AAAS archives; MacDougal to Cattell, 11 February 1914, MacDougal papers; "The American Association for the Advancement of Science," *Science* 33 (28 April 1911): 652; Robert C. Miller, "The AAAS on the Pacific Slope," *Science* 108 (3 September 1948): 220–223.

42. Cattell to MacDougal, 14 January 1913, 28 January 1913, 15 February 1913, 1 March 1913, 3 March 1913, 28 May 1913, 8 March 1914, 19 March 1914, 6 June 1914, 3 November 1914, MacDougal to Cattell, 22 January 1913, 11 February 1913, 25 February 1913, 19 March 1913, 11 February 1914, 17 February 1914, 5 March 1914, 13 March 1914, 1 April 1914, 13 May 1914, 16 June 1914, 3 November 1914, MacDougal papers; Cattell to Albert L. Barrows, 3 March 1914, 24 March 1914, 9 January 1915, 19 June 1915, Barrows to Cattell, 23 June 1914, 7 October 1914, 29 November 1914, 19 May 1915, 10 June 1915 (2 letters), Pacific Division AAAS archives, California Academy of Sciences, San Francisco; Minutes of the [AAAS] Committee on Policy, 17 November 1913, 28 December 1913, 29 December 1913, 20 April 1914, 27 December 1914, Minutes of the [AAAS] Council, 30 December 1913, 31 December 1913, 21 April 1914, 30 December 1914, AAAS archives.

43. "The Committee on the Pacific Coast Meeting of the American Association," *Science* 37 (2 May 1913): 660–661; Albert L. Barrows, "The Preliminary Announcement of the San Francisco Meeting of the American Association for the Advancement of Science," *Science* 39 (29 May 1914): 781–782; "The San Francisco Meeting of the American Association for the Advancement of Science,"

Science 40 (25 December 1914): 928–929; "Mathematics, Astronomy and Physics at the California Meeting," *Science* 41 (22 January 1915): 127; "The Pacific Division of the American Association," *Science* 41 (11 June 1915): 857; Pacific Coast issue *of Popular Science Monthly* 86 (March 1915), pp. 209–305.

44. Frank E. E. Germann, "The Southwestern Division of the AAAS," *Science* 108 (3 September 1948): 224–226; MacDougal to Livingston, 3 April 1920, MacDougal papers.

45. Howard C. Plotkin, "Edward C. Pickering and the Endowment of Scientific Research in America, 1877–1918," *Isis* 69 (1978): 44–57; Robert E. Kohler, *Partners in Science: Foundations and Natural Scientists 1900–1945* (Chicago: University of Chicago Press, 1991), 73–82; Nathan Reingold, "The Case of the Disappearing Laboratory," *American Quarterly* 29 (1977): 79–101; Nathan Reingold and Ida H. Reingold, eds., *Science in America: A Documentary History, 1900–1939* (Chicago: University of Chicago Press, 1981), pp. 193–197.

46. Minutes of the [AAAS] Committee on policy, 1 January 1914, Minutes of the [AAAS] Council, 2 January 1914, AAAS archives; Springsteen, "The American Association for the Advancement of Science: The Atlanta Meeting"; Edmund B. Wilson to Charles W. Eliot, 18 January 1914, Cattell to Edward C. Pickering, 21 February 1914, Cattell to Eliot, 3 March 1914, 19 March 1914, Charles W. Eliot presidential papers, Harvard University Archives, Cambridge; Cattell form letter to Committee on One Hundred members, 21 March 1914, Fairchild papers, MacDougal papers.

47. MacDougal to Cattell, 2 April 1914, MacDougal papers; William M. Davis to Cattell, 26 May 1914, Cattell papers; "The Committee of One Hundred on Scientific Research of the American Association for the Advancement of Science," *Science* 39 (8 May 1914): 680–682; "The Committee of One Hundred on Scientific Research," *Science* 40 (11 December 1914): 846; "The Committee on One Hundred on Scientific Research of the American Association for the Advancement of Science," *Science* 41 (26 February 1915): 315–320; Pickering to Cattell, 28 September 1914, Cattell to Pickering, 1 October 1914, Harvard College Observatory records, Harvard University Archives; Cattell to Pickering, 9 Januar 1915, Pickering papers.

48. Willis R. Whitney to George Ellery Hale, 9 January 1916, Hale papers, California Institute of Technology Archives, Pasadena.

49. Plotkin, "Edward C. Pickering and the Endowment of Scientific Research in America, 1877–1918"; Kohler, *Partners in Science*; Reingold, "The Case of the Disappearing Laboratory"; Reingold and Reingold, *Science in America*, pp. 244–246, 249–252.

50. "Report of the Pacific Coast Subcommittee on the Committee of One Hundred on Research," *Science* 43 (31 March 1916): 457–458.

51. Charles R. Cross, "Report of the Subcommittee on Research Funds," *Science* 41 (26 February 1915): 317–319; Cross, "Grants for Scientific Research," *Science* 43 (12 April 1916): 680–681; Cross, "Grants for Scientific Research: Medical Schools and Laboratories," *Science* 44 (14 July 1916): 50–51; Cross, "Grants for Scientific Research: Chiefly Collegiate Institution," *Science* 44 (18 August 1916): 229–232.

52. Minutes of the [AAAS] Committee on Policy, 20 April 1914, 31 December 1914, 19 April 1915, 17 April 1916, 30 September 1916, 29 December 1916, 22 April 1918, and Minutes of the [AAAS] Council, 29 December 1916, AAAS archives; Cross, "Grants for Scientific Research"; Cattell, "The Committee of One Hundred on Scientific Research of the American Association for the Advancement of Science," *Science* 45 (19 January 1917): 57–58; AAAS Committee on Grants for Research, Rules of Procedure, Adopted April 1917, Fairchild papers; Pickering to Cattell, 3 October 1916, 21 October 1916, Cattell to Pickering, 25 October 1916, 28 October 1916, 18 December 1916, Harvard College Observatory records; "The American Association for the Advancement of Science: Report of the Committee on Grants for Research," *Science* 45 (11 May 1917): 452–455.

53. George Ellery Hale, "The National Research Council: How It Proposes to Mobilize the Nation's Science for Industrial Progress and Military Efficiency," *New York Times*, 29 July 1916, p. 8; Dupree, *Science in the Federal Government*, pp. 308–315; Daniel J. Kevles, "George Ellery Hale, the First World War, and the Advancement of Science in America," *Isis* 59 (1968): 427–437; Rexmond C. Cochrane, The *National Academy of Sciences: The First Hundred Years, 1863–1963* (Washington: National Academy of Sciences, 1978), 208–216; Reingold, "The Case of the Disappearing Laboratory"; Kohler, *Partners in Science*, pp. 74–75.

54. Daniel J. Kevles, "Federal Legislation for Engineering Experiment Stations: The Episode of World War I," *Technology and Culture* 12 (1971): 182–189; "Resolutions of the Committee of One Hundred," *Scientific Monthly* 3 (1916): 99.

55. George Ellery Hale, "The National Research Council of the United States," *Nature* 97 (3 August 1916): 464–465; *Proceedings of the AAAS* 10 (1856): 239.

56. "The National Research Council," *Science* 44 (20 October 1916): 559–562; Minutes, Meeting of the [NRC] Executive Committee, 21 September 1916, Relations with AAAS (1916–1918) files, National Academy of Sciences archives, Washington, D.C.; Minutes of the [AAAS] Committee on Policy, 30 September 1916, AAAS archives; Howard to Hale, 3 October 1916, 9 October 1916, Hale papers.

57. Hale to Cary T. Hutchinson (NRC secretary), 5 December 1916, "Votes passed December 26, 1916, by the Committee of One Hundred on Research of the American Association for the Advancement of Science," Relations with AAAS (1916–1918) files; William M. Davis to Hale, 26 December 1916 (telegram), 30 December 1916, Hale papers.

58. Cattell to Fairchild, 10 July 1916, 6 November 1916, "Notes on Conference of the Committee . . . to consider plans for reorganization," 1 October 1916, MacDougal to Fairchild, 9 November 1916, 1 December 1916, Fairchild papers; Cattell to MacDougal, 10 July 1916, 26 July 1916, 28 October 1916, MacDougal to Cattell, 18 July 1916, 24 August 1916, 6 November 1916, Fairchild to MacDougal, 3 November 1916, 25 November 1916, undated draft "Report and Recommendation of the Sub-Committee on the Federation of Scientific Organizations in America," MacDougal papers; Minutes of the [AAAS] Committee on Policy, 30 September 1916, 26 December 1916, 27 December 1916, AAAS archives.

59. "Suggestions for Final Report of the Committee," 1 October 1916, "Minutes of the Meeting Held at the Century Club," 9 December 1916, MacDougal to Fairchild, 20 December 1916, Fairchild papers; Fairchild to MacDougal, 10 December 1916, 22 December 1916, Humphreys to MacDougal, 19 December 1916, Cattell to MacDougal, 20 December 1916, 22 December 1916, 25 January 1917, 7 February 1917, MacDougal to Cattell, 21 December 1916, 8 January 1917, 1 February 1917, 13 February 1917, MacDougal papers.

60. Minutes of the [AAAS] Council, 1 January 1918, AAAS archives.

61. "The Pittsburgh Meeting of the American Association for the Advancement of Science," *Scientific Monthly* 6 (1918): 91–92, 189–191; MacDougal to Nathaniel L. Britton, 9 January 1918, MacDougal papers.

62. Carol Singer Gruber, *Mars and Minerva: World War I and the Uses of the Higher Learning in America* (Baton Rouge: Louisiana State University Press, 1976), pp. 188–206; Woodward to Henry F. Osborn, 19 October 1917, Osborn papers, American Museum of Natural History, New York; Howard to Cattell, 10 November 1917, Cattell papers.

63. Minutes of the [AAAS] committee on policy, 19 November 1917, 22 April 1918, 13 November 1919, Minutes of the [AAAS] Council, 29 December 1917, 30 December 1917, AAAS archives.

64. Kohler, *Partners in Science.*

65. "The Baltimore Meeting of the American Association for the Advancement of Science," *Scientific Monthly* 7 (1918): 571–573; "Convocation-Week Meeting of the American Association for the Advancement of Science," *Scientific Monthly* 8 (1919): 87–93; MacDougal to Cattell, 26 July 1920, Cattell papers.

66. "The Secretaryship of the American Association for the Advancement of Science," *Science* 50 (28 November 1919): 498–499; Sam F. Trelease, "The Work of Dr. Burton E. Livingston as Permanent Secretary of the American Association," *Scientific Monthly* 32 (1931): 281–283; Charles A. Shull, "Burton Edward Livingston, 1875–1948," *Science* 107 (28 May 1948): 558–560.

67. Cattell to A. A. Noyes, 12 January 1920, Simon Flexner papers, American Philosophical Society Library, Philadelphia; Minutes of the [AAAS] Executive Committee, 17 October 1920, Sam Woodley to Burton E. Livingston, 26 January 1944, AAAS archives; Burton E. Livingston, "Sam Woodley: Executive Assistant of the American Association for the Advancement of Science," *Science* 101 (1 June 1945): 551–552; Livingston, "Present Status of the Affairs of the American Association for the Advancement of Science," *Science* 52 (24 December 1920): 599–600.

68. Robert E. Baker, "Changes in the Constitution of the American Association for the Advancement of Science: Late 1890s, ca. 1920, and ca. 1945," unpublished undergraduate paper, Worcester Polytechnic Institute, Worcester, Mass., May 1983.

69. Minutes of the [AAAS] Council, 1 January 1920, 28 December 1920, 30 December 1920, Minutes of the [AAAS] Executive Committee, 28 December 1925, AAAS archives.

70. Livingston to MacDougal, 15 August 1922, 19 November 1922, MacDougal papers.

71. Minutes of the [AAAS] Executive Committee, 25 October 1931, AAAS archives.
72. Wolfle, *Renewing a Scientific Society*, pp. 135–137.
73. Minutes of the [AAAS] Executive Committee, 26 April 1920, 26 April 1925, AAAS archives.
74. Maynard M. Metcalf, "The Scientific Spirit," *Science* 49 (13 June 1919): 551–558; Howard to Cattell, 8 July 1919, Cattell papers; George T. Moore, "AAAS: Report of the St. Louis Meeting," *Science* 51 (9 January 1920): 48–50; "Scientists Rebuke Political Partisans," *St. Louis Times*, 3 January 1920.
75. Minutes of the [AAAS] Executive Committee, 24 April 1927, AAAS archives.
76. Cattell, "Report of the Committee on Convocation Week," *Science* 56 (1 December 1922): 616–620; "The Salt Lake City Meeting," *Scientific Monthly* 14 (1922): 587–588.
77. "The Scientific Meetings at Chicago," *Scientific Monthly* 11 (1920): 569–570; Dorothy M. Livingston, *The Master of Light: A Biography of Albert A. Michelson* (Chicago: University of Chicago Press, 1973), p. 277; Arthur H. Compton, "Personal Reminiscences," in *The Cosmos of Arthur Holly Compton*, ed. Marjorie Johnston (New York: Alfred A. Knopf, 1967), pp. 3–52.
78. Kenneth R. Manning, *Black Apollo of Science: The Life of Ernest Everett Just* (New York: Oxford University Press, 1983), pp. 84–87.
79. Minutes of the [AAAS] Council, 1 January 1926, AAAS archives.
80. Frederick E. Brasch, "The History of Science and the American Association for the Advancement of Science," *Science* 50 (18 July 1919): 66–68; Florian Cajori, "The History of Science and the American Association for the Advancement of Science," *Science* 53 (18 February 1921): 163–164; Livingston to MacDougal, 21 July 1922, MacDougal papers.
81. Livingston to MacDougal, 17 April 1922, 21 July 1922, MacDougal papers.
82. Livingston to MacDougal, 26 July 1922, MacDougal papers; Michael M. Sokal, "Companions in Zealous Research, 1886–1986," *American Scientist* 74 (1986): 486–508.
83. Livingston, "Permanent Secretary's Annual Report on the Affairs of the Association," 25 October 1925; Minutes of the [AAAS] Executive Committee, 25 October 1925, AAAS archives.
84. Minutes of the [AAAS] Executive Committee, 1 January 1925, 25 October 1925, 25 April 1926, 30 November 1930, 29 December 1930, AAAS archives; Livingston, "The Annual Science Exhibition of the American Association," *Science* 64 (19 November 1926): 491–492; Livingston, "Report of the Permanent Secretary on Association Affairs," 1 April 1930, AAAS archives.
85. Sokal, "*Science* and James McKeen Cattell"; Albert Einstein, "Time, Space, and Gravitation," *Science* 51 (2 January 1920): 8–10; Clinton J. Davisson and C. H. Kunsman, "The Scattering of Electrons by Nickel," *Science* 54 (25 November 1921): 522–524; Davisson and Lester H. Germer, "Diffraction of Electrons by a Crystal of Nickel," *Physical Review* 30 (1927): 705–740.
86. Livingston to MacDougal, 21 July 1922, 26 July 1922, 9 August 1922, MacDougal to Livingston, 4 August 1922, MacDougal papers.
87. Livingston to MacDougal, 5 February 1922, Livingston to Cattell, 5 November 1922, MacDougal papers.

88. MacDougal to Cattell, 21 December 1921, W. J. Humphreys to MacDougal, 23 December 1921, Cattell papers; MacDougal, "Publications of the American Association for the Advancement of Science," December 1921, AAAS archives; Minutes of the [AAAS] Executive Committee, 30 October 1921, 21 October 1922, AAAS archives.

89. Livingston to MacDougal, 25 July 1922, 26 July 1922, MacDougal to Livingston, 4 August 1922, MacDougal papers.

90. Minutes of the [AAAS] Executive Committee, 27 December 1922, 22 April 1923, AAAS archives; Livingston to MacDougal, 3 January 1923, 24 January 1923, 17 July 1923, 3 April 1924, MacDougal to Livingston, 12 January 1923, 29 January 1923, 25 July 1923, MacDougal papers.

91. Sokal, "*Science* and James McKeen Cattell"; Humphreys to Cattell, 16 April 1925, Cattell to Humphreys, 21 April 1925, Livingston to Cattell, 12 May 1925, Cattell to Livingston, 16 May 1925, Cattell papers; Livingston to Roscoe Pound, 19 May 1925, Pound to Livingston, 20 May 1925, 25 June 1925, Cattell to Pound, 25 May 1925, Livingston to Cattell, 11 July 1925, Minutes of the [AAAS] Executive Committee, 26 April 1925, 28 December 1925, Minutes of the [AAAS] Council, 28 December 1925, AAAS archives; Cattell, "The Journal 'Science' and the American Association for the Advancement of Science," *Science* 64 (8 October 1926): 342–347.

92. Livingston to MacDougal, 14 November 1925, MacDougal papers.

93. Livingston to MacDougal, 25 July 1922, MacDougal papers; Ronald C. Tobey, *The American Ideology of National Science*, pp. 63–66; David J. Rhees, "A New Voice for Science: Science Service under Edwin E. Slosson, 1921–29," M.A. thesis, University of North Carolina, 1979; Daniel J. Kevles, *The Physicists: The History of a Scientific Community in Modern America* (New York: Alfred A. Knopf, 1978), pp. 171, 174.

94. Minutes of the [AAAS] Council, 31 December 1925, 1 January 1930, Minutes of the [AAAS] Executive Committee, 25 April 1926, 24 April 1927, 28 December 1928, 20 October 1929, AAAS archives; Livingston, "The American Association for the Advancement of Science: Newspaper Reports on the Meetings," *Science* 66 (21 October 1927): 368–370; Austin H. Clark, "Press Service," *Science* 69 (22 February 1929): 219–220.

95. "Glimpses of the March of Science," *Literary Digest*, 12 January 1929, p. 10; James J. Walsh, "The Scientists Convene," *Commonweal* 9 (16 January 1929): 315–317; "Important Discussions at Des Moines Science Meeting," *Catholic World* 130 (1930): 620–621; Clark, "The Work of the Press," *Science* 79 (9 February 1934): 141–142; Clark, "The Press Service at the Pittsburgh Meeting," *Science* 81 (29 March 1935): 315–316; "Broadcasts by the American Association for the Advancement of Science," *Science* 84 (25 December 1936): 570.

96. Livingston to MacDougal, 21 July 1922, MacDougal papers; Minutes of the [AAAS] Executive Committee, 24 April 1927, AAAS archives.

97. Minutes of the [AAAS] Council, 27 December 1929, Minutes of the [AAAS] Executive Committee, 29 December 1929, AAAS archives; Otis W. Caldwell, "The American Association's Committee on the Place of Science in Education," *Science* 60 (24 October 1924): 376–377; Caldwell, "American Association for

the Advancement of Science Committee on the Place of the Sciences in Education," *Science* 60 (12 December 1924): 536–540; Joseph L. Wheeler, "American Association Science Booklists," *Library Journal* 56 (15 December 1931): 1062–1064; Morris Meister, "Otis William Campbell, 1869–1947," *Science* 106 (12 December 1947): 576–578; Wolfle, *Renewing a Scientific Society*, pp. 149–152.

98. For example, Minutes of the [AAAS] Executive Committee, 27 December 1920, 24 April 1921, Minutes of the [AAAS] Council, 29 December 1920, 27 December 1921, AAAS archives; "Duty on Scientific Apparatus for Educational Institutions," *Science* 53 (27 May 1921): 494–495; Gayner Glass Works to Cattell, 21 June 1921, Cattell papers; Providence Chemical Laboratories to MacDougal, 30 June 1921, MacDougal papers.

99. Minutes of the [AAAS] Executive Committee, 26 April 1925, 25 October 1925, "Permanent Secretary's Semi-Annual Report on the Affairs of the Association," 26 April 1925; "Permanent Secretary's Annual Report on the Affairs of the Association," 25 October 1925; AAAS archives; "The John T. Scopes Scholarship Fund," *Science* 62 (31 July 1925): 105; "Scientific Men and the Defense of Mr. Scopes," *Science* 62 (7 August 1925): 130; "The Scopes Scholarship Fund," *Science* 62 (25 September 1925): 282; Ray Ginger, *Six Days or Forever: Tennessee vs. John Thomas Scopes* (Boston: Beacon Press, 1958), pp. 77, 81, 125–126, 202; Edward J. Larson, *Summer of the Gods: The Scopes Trial and America's Continuing Debate over Science and Religion* (New York: Harper Collins, 1997), pp. 113–116, 134–135, 210–211; Samuel Walker, *In Defense of American Liberty: A History of the American Civil Liberties Union* (New York: Oxford University Press, 1990), pp. 75–76.

100. Newcomb Cleveland to Josephine Owen Cattell, 27 January 1944, Cattell papers; Livingston, "The AAAS: The Permanent Secretary's Report of the Cincinnati Meeting," *Science* 59 (25 January 1924): 71–79; The Second Annual American Association Prize," *Science* 61 (13 March 1925): 277–279; Minutes of the [AAAS] Executive Committee, 27 April 1924, 24 April 1927, AAAS archives; Kevles, *The Physicists*, p. 227; Livingston, *The Master of Light*, p. 314.

101. Appel, "Organizing Biology," pp. 110–112; Livingston, "Permanent Secretary's Annual Report on the Affairs of the Association, 18 October 1922, AAAS archives; Minutes of the [AAAS] Executive Committee, 23 April 1922, 21 October 1922, 23 April 1923, 27 December 1923, 30 December 1924, Minutes of the [AAAS] Council, 28 December 1922, 30 December 1924, 31 December 1926, 28 December 1928, AAAS archives.

102. Ellen Condliffe Lagemann, *The Politics of Knowledge: The Carnegie Corporation, Philanthropy, and Public Policy* (Middletown, Conn.: Wesleyan University Press, 1989), pp. 43–50; Cattell to MacDougal, 2 April 1920, MacDougal to Cattell, 7 April 1920, MacDougal papers.

103. Livingston, "Relations of the American Association to the National Academy," *Science* 66 (25 November 1927): 493–495.

104. Cattell, "The Organization of Scientific Men," *Scientific Monthly* 14 (1922): 567–577; undated galley proofs, Cattell to Hale, 4 May 1920, 29 May 1920, James R. Angell to Cattell, 20 May 1920, Cattell papers; Hale to Cattell, 14

May 1920, 3 June 1920, Hale papers; Nathan Reingold, "National Aspirations and Local Purposes," *Transactions of the Kansas Academy of Sciences* 71 (1968): 235–246.

105. For example, Edwin B. Wilson to Henry B. Ward, 25 July 1932, 28 September 1932, 10 March 1934, 13 November 1934, 30 November 1934, Wilson to Charles F. Roos, 22 September 1932, 24 October 1932, Ward to Wilson, 8 March 1934, Edwin B. Wilson papers, Harvard University Archives, Cambridge.

106. Livingston to Fairchild, 17 February 1931, Fairchild papers; Livingston to Edwin B. Wilson, 17 February 1930, 24 February 1930, 4 November 1930, Wilson to Roos, 3 May 1932, 7 October 1932, Roos to Wilson, 18 September 1932, 27 September 1932, 21 October 1932, Edwin B. Wilson papers; Cattell to Livingston, 6 December 1930, Cattell to Roos, 17 September 1932, Cattell to Ward, 6 October 1932, Cattell papers.

107. Edwin B. Wilson to Ward, 22 August 1932, 19 September 1932, 10 March 1934, Edwin B. Wilson papers; Cattell to Ward, 26 September 1932, Cattell papers.

108. Edwin B. Wilson to Ward, 25 July 1932, Wilson to Roos, 22 September 1932, Edwin B. Wilson papers.

109. Cattell to Livingston, 19 June 1930, Cattell papers; Edwin B. Wilson to Ward, 22 August 1932, 28 September 1932, 10 March 1934, 30 November 1934, Wilson to Roos, 22 September 1932, 7 March 1934, Edwin B. Wilson papers; Report of the Committee on Organization, 26 December 1934, Burton E. Livingston papers, Rutgers University, New Brunswick, NJ.

110. Sokal, "Companions in Zealous Research," p. 494; Minutes of the [AAAS] Executive Committee, 27–30 December 1936, AAAS archives; Anton J. Carlson, "Forest Ray Moulton, 1872–1952," *Science* 117 (22 May 1953): 545–546.

111. Kevles, *The Physicists*, pp. 241–242; Robert H. Kargon, *The Rise of Robert Millikan: Portrait of a Life in American Science* (Ithaca: Cornell University Press, 1982), pp. 156–158.

112. Minutes of the [AAAS] Executive Committee, 29 December 1930, 24 April 1932, 15–16 October 1932, 27 December 1932, AAAS archives; "Foreign Guests at the Century of Progress Meeting of the American Association for the Advancement of Science," *Science* 76 (25 November 1932): 484–485; John L. Heilbron and Robert W. Seidel, *Lawrence and His Laboratory: A History of the Lawrence Berkeley Laboratory*, vol. 1 (Berkeley: University of California Press, 1989), pp. 156–157; William L. Laurence, "New 'Gun' Speeds the Breakup of Atom," *New York Times*, 20 June 1933, p. 1; "Complementarity in Chicago," *Time* (3 July 1933): 40.

113. "Some Scientific Facts, Some Faces of Scientists Seen in Atlantic City," *Life* (16 January 1937): 35–38; James J. Walsh, "The Scientists Convene," *Commonweal*; "Important Discussions at Des Moines Science Meeting," *Catholic World*; "Scientists Are Quite Sure of Themselves," *Christian Century* 50 (5 July 1933): 868–869; "Harvard Professor Explores 'Reverent Science,'" *Literary Digest* 117 (6 January 1934): 8.

114. Minutes of the [AAAS] Executive Committee, 30–31 October 1937, 24 April 1938, 22–23 October 1938, 26–30 December 1938, 23 April 1939, 27–30 De-

cember 1939, 28 January 1945, 11 March 1945, AAAS archives; Moulton, "Science by Radio," *Scientific Monthly* 47 (1938) 546–548.

115. Minutes of the [AAAS] Executive Committee, 25–26 April 1931, 15–16 October 1933, 14 April 1935, 26 December 1937, AAAS archives; F. C. Brown, "The Annual Science Exhibition," *Science* 86 (27 August 1937): 191–192; Brown, "Film Showings at the Annual Science Exhibition," *Science* 86 (12 November 1937): 434; "Annual Exhibition of the American Association for the Advancement of Science and Associated Societies," Philadelphia, 27–31 December 1940, Walter R. Miles papers, Archives of the History of American Psychology, University of Akron, Akron, Ohio.

116. Cattell, "Book Reviews in Science," *Science* 69 (22 February 1929): 220–222; Albert Einstein, "Lens-Like Action of a Star by the Deviation of Light in the Gravitational Field," *Science* 84 (4 December 1936): 506–507; Cattell to Alexis Carrel, 28 February 1931, Alexis Carrel papers, Georgetown University Library, Washington, D.C.; Cattell to John J. Abel, 9 December 1931, John J. Abel papers, Johns Hopkins University Medical Archives, Baltimore; Cattell to George S. Myers, 10 November 1934, Division of Fisheries Records, Smithsonian Institution Archives, Washington, D.C.; E. E. Just to Cattell, 25 November 1929, Hudson Hoagland to Cattell, 7 June 1937, Cattell to Thomas H. Morgan, 10 December 1938, Cattell papers.

117. For example, Ward to D. H. Palmer, 30 March 1935, Raymond Pearl to Cattell, 22 April 1938, Leonard Carmichael to Cattell, 25 July 1938, Cattell papers; Moulton to Bela Hubbard, 1 March 1939, William G. Clark to Moulton, 20 July 1942, AAAS archives.

118. Moulton to William G. Clark, 28 July 1942, Minutes of the [AAAS] Executive Committee, 15–16 October 1933, AAAS archives; Melvin J. Herskovitz, "Pavlov's Bequest," *New Republic*, 20 January 1937, p. 360.

119. Minutes of the [AAAS] Executive Committee, 27 December 1936, 30 December 1936, 17 April 1937, 20 June 1937, 26 December 1937, 24 April 1938, 22–23 October 1938, AAAS archives; Livingston to Woodley, 27 March 1937, Livingston to Edwin G. Conklin, 31 March 1937, Edwin G. Conklin papers, Princeton University Library, Princeton, NJ; Cattell to Moulton, 4 December 1937, 10 December 1937, 18 December 1937, 22 September 1938, 6 October 1938, 13 October 1938, 28 October 1938, 2 November 1938, 12 November 1938, 22 November 1938, 26 November 1938, Moulton to Cattell, 8 December 1937, 14 December 1937, 21 December 1937, 6 June 1938, 24 September 1938, 1 October 1938, 10 October 1938, 17 October 1938, 31 October 1938, 4 November 1938, 7 November 1938, 9 November 1938, 14 November 1938, 24 November 1938, 28 November 1938, Cattell papers.

120. Minutes of the [AAAS] Executive Committee, 16 April 1938, Conklin papers; Moulton, "The Second Denver Meeting of the American Association for the Advancement of Science," *Science* 86 (13 August 1937): 133–134; Cattell, "The Scientific Monthly and the American Association for the Advancement of Science," *Science* 88 (1938): 428–429.

121. Kevles, *The Physicists*, p. 268; Heilbron and Seidel, *Lawrence and His Laboratory*, p. 104.

122. Minutes of the [AAAS] Executive Committee, 15–16 October 1932, 24–25 March 1933, 15–16 October 1933, 28 December 1933, 24–25 October 1935, 27 October 1935, 17 April 1937, 26 December 1937, AAAS archives.

123. Minutes of the [AAAS] Executive Committee, 25 October 1931, 26–27 December 1933, 15 April 1934, 27 December 1934, 31 December 1934, AAAS archives; "Curtailment of Scientific Work Under the Government," *Science* 77 (23 June 1933): 600; "Proposed Grants for Scientific Research from the Public Works Board," *Science* 78 (14 July 1933): 30–31; "Science and Public Works," *Science* 78 (21 July 1933): 61; "The Scientific Work of the Government," *Science* 78 (11 August 1933): 122–123; "The Financial Situation of the Department of Agriculture," *Science* 78 (18 August 1933): 140; "Curtailment of the Federal Scientific Bureaus," *Science* 78 (25 August 1933): 162–163; "Federal Allotment for Public Works," *Science* 78 (15 September 1933): 240–241.

124. Ward to the Civil Works Administration, 13 December 1933, 26 January 1934, Ward to Cattell, 26 January 1934, 8 February 1934, 17 February 1934, 15 May 1934, Cattell papers; Minutes of the [AAAS] Executive Committee, 27 December 1934, 31 December 1934; AAAS archives; Dupree, *Science in the Federal Government*, pp. 350–358.

125. "Italian Professors and the Fascist Government," *Science* 75 (15 January 1932): 73; "Academic Freedom in Italy," *Science* 75 (25 March 1932): 338; "The Situation of Jewish Scientific Men and Physicians in Germany," *Science* 77 (19 May 1933): 492–494; "The Scientific Situation in Germany," *Science* 77 (2 June 1933): 528–529; "The British Academic Council," *Science* 77 (30 June 1933): 621; "Academic Assistance," *Science* 77 (30 June 1933): 629; "Committee of Dutch Professors on Behalf of German Jewish Students and Graduates," *Science* 78 (7 July 1933): 7–8; "The Emergency Committee in Aid of Displaced German Scholars," *Science* 78 (21 July 1933): 52–53; "Academic Freedm in Germany," *Science* 78 (17 November 1933): 460–461; "The Disgrace of German Science," *Science* 79 (27 April 1934): 386–387; "Intellectual Freedom," *Science* 87 (7 January 1938): 10.

126. Robert E. Filner, "The Roots of Political Activism in British Science," *Bulletin of the Atomic Scientists*, January 1976, pp. 25–29; Filner, "The Social Relations of Science Movement (SRS) and J.B.S. Haldane," *Science & Society* 41 (1977): 303–316; William McGucken, "The Social Relations of Science: The British Association for the Advancement of Science, 1931–1946," *Proceedings of the American Philosophical Society* 123 (1979): 236–264; McGucken, *Scientists, Society, and State: The Social Relations of Science Movement in Great Britain, 1931–1947* (Columbus: Ohio State University Press, 1984); Gary Werskey, *The Visible College: The Collective Biography of British Scientific Socialists in the 1930s* (New York: Holt, Rinehart and Winston, 1979).

127. Dugald C. Jackson to Cattell, 17 June 1935, 10 July 1935, 26 July 1935, Cattell to Jackson, 21 June 1935, 23 July 1935, Dugald C. Jackson papers, Massachusetts Institute of Technology archives, Cambridge, Mass.; Jackson, "Objectives of Engineering Education," *Journal of Engineering Education* 26 (1935): 60–85.

128. Peter J. Kuznick, *Beyond the Laboratory: Scientists and Political Activists in*

1930s America (Chicago: University of Chicago Press, 1987), pp. 68–73; Minutes of the [AAAS] Executive Committee, 17 April 1937, 20 June 1937, AAAS archives.

129. Kuznick, *Beyond the Laboratory*, pp. 227–252; Minutes of the [AAAS] Executive Committee, 26–31 December 1940, 18 October 1942, 17 January 1943, AAAS archives.

130. Moulton, "American Association for the Advancement of Science," unpublished typescript, 17 November 1937, Miles papers.

131. Kuznick, *Beyond the Laboratory*, pp. 74–76; "Science and Democracy," *New York Times*, 17 October 1938, section IV, p. 8; "Science and Democracy," *Science* 86 (22 October 1938): 375–376.

132. Minutes of the [AAAS] Council, 30 December 1937, AAAS archives; Kuznick, *Beyond the Laboratory*, pp. 76–82.

133. Kuznick, *Beyond the Laboratory*; Conklin, "Science and Ethics," *Science* 86 (31 December 1937): 595–603; Moulton, "The Spirit of It," *Science* 87 (4 February 1938): 95–96.

134. Minutes of the [AAAS] Executive Committee, 26 December 1937, 24 April 1938, 26–30 June 1938, AAAS archives; Kuznick, *Beyond the Laboratory*, pp. 88–90; "Cooperation Between the British and American Associations," *Science* 88 (16 September 1938): 258–259; Moulton, "The British and American Associations," *Science* 88 (23 September 1938): 276–277; Moulton, "Science and War," *Science* 88 (7 October 1938): 324–325.

135. Kuznick, *Beyond the Laboratory*, pp. 96–105; William L. Laurence, "Scientists Gird to Rescue World from Misuse of Man's Inventions," *New York Times*, 26 December 1938, pp. 1, 20; Richard Gregory, "Science, Religion and Social Ethics," *Science* 89 (24 February 1939): 163–166; Waldemar Kaempffert, "The Week in Science," *New York Times*, 1 January 1939, section II, p. 11; "Physicist Pictures New 'Free Society,'" *New York Times*, 30 December 1938, p. 2; Percy W. Bridgman, "'Manifesto' by a Physicist," *Science* 89 (24 February 1939): 179.

136. Minutes of the [AAAS] Executive Committee, 28 December 1931, 28 December 1932, 15 April 1934, 21 October 1934, 19 April 1936, 30–31 October 1937, 26 December 1937, 24 April 1938, 26–30 June 1938, 22–23 October 1938, 26–30 December 1938, 19–21 June 1939, AAAS archives; Otis W. Caldwell, "Work of the Committee on the Place of Science in Education," *School and Society* 40 (24 November 1934): 673–679; "The Committee on the Improvement of Science in General Education," *Science* 87 (20 May 1938): 454–455; "Grants-in-Aid for Studies in Science Instruction," *Science* 88 (23 December 1938): 588–589; Lloyd W. Taylor, "Improvement of Science Instruction," *Science* 89 (13 January 1939): 34.

137. Minutes of the [AAAS] Executive Committee, 15 April 1934, 21 October 1934, 2 January 1936, 19 April 1936, AAAS archives; Harlow Shapley to members of the AAAS Committee on Adult Education, 21 June 1934, Harvard College Observatory records.

138. Carlyle B. Storm and Jimmie C. Oxley, *Gordon Research Conferences: 50 Years in New Hampshire* (Kingston, R.I.: Gordon Research Conferences, 1997).

139. "Local Branches of the American Association for the Advancement of Science,"

Science 39 (13 February 1914): 246–247; Cattell, "Local Branches of the American Association for the Advancement of Science," *Science* 80 (21 December 1934): 576–578; "The Lancaster Branch of the American Association for the Advancement of Science," *Science* 81 (22 March 1935): 286–287; Minutes of the [AAAS] Committee on Policy, 17 November 1913, 29 December 1913, 29 December 1916, AAAS archives; Minutes of the [AAAS] Executive Committee, 14 April 1935, AAAS archives.

140. Moulton, "The American Association—An Integrating Agency in Science," *Scientific Monthly* 54 (1942): 186–188; Hudson Hoagland, "Analysis of Post-War Problems and Procedures," *Science* 95 (20 February 1942): 195–196; Gustav Egloff, "Peacetime Values from a War Technology," *Science* 97 (29 January 1943): 101–108; Lawrence K. Frank, "Research after the War," *Science* 101 (27 April 1945): 433–434; Theodore Rosebury, "The Fuller Utilization of Scientific Resources for Total War," *Science* 96 (25 December 1942): 571–575; K.A.C. Elliott and Harry Grundfest, "The Science Mobilization Bill," *Science* 97 (23 April 1943): 375–377; Harley M. Kilgore, "The Mobilization of Science," *Science* 97 (7 May 1943): 407–412; John Q. Stewart, "The 'Science Mobilization Bill,'" *Science* 97 (28 May 1943): 486–487; Robert F. Maddox, "The Politics of World War II Science: Senator Harley M. Kilgore and the Legislative Origins of the National Science Foundation," *West Virginia History* 41 (1979): 20–39; Daniel J. Kevles, "The National Science Foundation and the Debate over Postwar Research Policy, 1942–1945," *Isis* 68 (1977): 5–26; Milton Lomask, *A Minor Miracle: An Informal History of the National Science Foundation* (Washington, D.C.: NSF, 1976), pp. 40–42, 49–53; J. Merton England, *A Patron for Pure Science: The National Science Foundation's Formative Years, 1945–57* (Washington, D.C.: NSF, 1982), pp. 3, 9–24.

141. Albert F. Blakeslee, "Report of President to Council of A₃S, Philadelphia, December 27, 1940," AAAS archives; Blakeslee, "Report of the President to Council of American Association for the Advancement of Science," *Science* 93 (7 March 1941): 219–227.

142. Charles S. Baker (AAAS attorney) to Blakeslee, 20 December 1940, Irving Langmuir to Miles, 20 May 1941, Minutes of the [AAAS] Committee on Publications, 8 September 1941, Miles papers; Ware Cattell to James McKeen Cattell, 24 December 1940, Esmond R. Long to Blakeslee, 22 January 1941, Cattell papers; Blakeslee to Vannevar Bush, 3 January 1941, Bush to Blakeslee, 14 January 1941, Vannevar Bush papers, Library of Congress, Washington, D.C.; Blakeslee to Walter B. Cannon, 17 January 1941, Walter B. Cannon papers, Countway Library of Medicine, Boston, MA; Blakeslee to Cattell, 20 February 1941, Blakeslee to Moulton, 24 February 1941, Moulton to Caldwell, 9 May 1941, Moulton to Langmuir, 20 June 1941, Harlow Shapley to Langmuir, 26 June 1941, Langmuir to Kirtley F. Mather, 13 August 1941, 20 August 1941, Mather to Langmuir, 18 August 1941, Moulton to Mather, 12 September 1941, 15 September 1941, AAAS archives; Mather to Shapley, 27 August 1941, Shapley to Mather, 28 August 1941, 9 September 1941, Mather to Members of the Committee on Publications, 3 December 1941, Shapley papers.

143. Proposed Resolution, [AAAS] Committee on Publications, November 1941,

Miles papers; Cattell to Moulton, 1 November 1941, 8 November 1941, Moulton to Cattell, 6 November 1941, Cattell papers; Minutes of the [AAAS] Committee on Publications, 3 November 1941, Shapley papers; Livingston to Cattell, 19 November 1941, 24 November 1941, Moulton to Livingston, 28 November 1941, Cattell to Livingston, 24 December 1941, Livingston papers.

144. Cattell to Moulton, 22 September 1938, 30 March 1942, Moulton to Cattell, 24 September 1938, Cattell to Livingston, 27 March 1942, Livingston to Moulton, 29 March 1942, 10 April 1942, 2 October 1942, 12 May 1943, Moulton to Livingston, 1 April 1942, 30 September 1942, Caldwell to Moulton, 23 June 1943, Warren E. Magee (AAAS attorney) to Livingston, 2 July 1943, 7 July 1943, Minutes of the [AAAS] Executive Committee, 18 October 1942, 17 January 1943, 25 June 1943, 28 January 1945, 11 March 1945, 22 April 1945, AAAS archives; Ware Cattell to Cannon, 23 September 1943, Cannon papers; Moulton "to the Council of the AAAS," 19 October 1943, Cattell papers; Civil Action no. 21,508, U.S. District Court, Washington, DC, filed 6 October 1943; "Editor's Suit Settled," *New York Times*, 26 January 1945, p. 19; "Ware Cattell vs. AAAS Settled for $7,500 by Consent Judgment," *Science* 102 (6 July 1945): 5–6.

145. Livingston to Caldwell, 20 January 1944, Caldwell to Livingston, 24 January 1944, Livingston to Carlson, 18 April 1944, Livingston papers; Moulton to Members of the [AAAS] Executive Committee, 9 February 1944, Cannon papers; Minutes of the [AAAS] Executive Committee, 6 February 1944, 7 May 1944, 6 August 1944, 24 June 1945, 2 September 1945, AAAS archives; AAAS Agreement with Josephine Owen Cattell and Jaques Cattell, 7 May 1944, AAAS archives; AAAS Auditor's Report for 1954, AAAS archives.

146. Minutes of the [AAAS] Executive Committee, 6 August 1944, 22 August 1944, 14 November 1944, 24 June 1945, AAAS archives; Charles S. Stephenson, "The Problem: To Publish *Science*," undated memorandum, AAAS archives; Stephenson to Cannon, 7 November 1944, Moulton to Stephenson, 13 November 1944, Cannon papers; Civil Action no. 28,471, U.S. District Court, Washington, D.C., filed 10 April 1945.

147. Minutes of the [AAAS] Executive Committee, 14 November 1944, 28 January 1945, 11 March 1945, 26 April 1945, 24 June 1945, 2 September 1945, AAAS archives; Moulton, "The Appointment of Willard L. Valentine as Editor of *Science*," *Science* 102 (19 October 1945): 387–388.

148. Minutes of the [AAAS] Executive Committee, 6 August 1944, 22 October 1944, 11 March 1945, 24 June 1945, 2 September 1945, AAAS archives; Wolfle, *Renewing a Scientific Society*, pp. 101–110.

149. Minutes of the [AAAS] Executive Committee, 22 April 1945, AAAS archives; Wolfle, *Renewing a Scientific Society*, pp. 5–9.

150. Minutes of the [AAAS] Executive Committee, 6 August 1944, 22 October 1944, 14 November 1944, 11 March 1945, AAAS archives; Kennard Baker Bork, "Kirtley Fletcher Mather as Geologist and Political Activist," *Journal of Geological Education* 37 (1989): 190–193; Bork, *Cracking Rocks and Defending Democracy—Kirtley Fletcher Mather: Scientist, Teacher, Social Activist, 1888–1978* (San Francisco: Pacific Division of the AAAS, 1994), pp. 183, 190, 226–231.

Shifting Science from People to Programs: AAAS in the Postwar Years

1. AAAS, *The Organization and Work of the American Association for the Advancement of Science* (Washington, D.C.: AAAS, 1930), p. 1.
2. Dael Wolfle, *Renewing a Scientific Society: The American Association for the Advancement of Science from World War II to 1970* (Washington, D.C.: AAAS, 1989), pp. 10, 43.
3. AAAS, *Summarized Proceedings, 1915–1921* (Washington, D.C.: AAAS, 1921), pp. 10, 19.
4. AAAS, *Summarized Proceedings and Directory, 1940–1948* (Washington, D.C.: AAAS, 1948), p. 23.
5. F. R. Moulton, "Constitutions of the Association," *AAAS Bulletin* 5, no. 4 (1946): 29–30.
6. Wolfle, *Renewing a Scientific Society*, pp. 13–17.
7. *AAAS Bulletin* 1, no. 1 (March 1942): 1.
8. Harlow Shapley, "Some Plans for the American Association for the Advancement of Science," 21 July 1947, AAAS archives.
9. Harlow Shapley to Kirtley Mather, 1 November 1951, AAAS executive office files, "Arden House" folder (see also, in general, the concerns expressed throughout this folder); and "An AAAS Television Series," AAAS board minutes, 3–4 March 1956, suppl. to item 25.
10. Wolfle, *Renewing a Scientific Society*, p. 47.
11. Robert E. Kohler, "Warren Weaver and the Rockefeller Foundation Program in Molecular Biology: A Case Study in the Management of Science," in *The Sciences in the American Context: New Perspectives*, ed. Nathan Reingold (Washington, D.C.: Smithsonian Institution Press, 1979), 249–293.
12. For an introduction to Weaver and his concerns, see: Warren Weaver, *Scenes of Change: A Lifetime in American Science* (New York: Scribner's, 1970); Mina Rees, "Warren Weaver," *National Academy of Sciences Biographical Memoirs* 57 (1987); "Arches of Science Award," *Understanding* 4, no. 2 (spring 1966): 1, 6 (AAAS Misc. Pub. 66–10); and Bruce V. Lewenstein, "'Public Understanding of Science' in America, 1945–1965" (Ph.D. diss., University of Pennsylvania, 1987), pp. 329–365. Among Weaver's own writings, the following items speak to his concerns about popular science: Weaver, ed., *The Scientists Speak* (New York: Boni and Gaer, 1947); Weaver, "Science and People," in *The Maturing of American Science*, ed. Robert Kargon (Washington, D.C.: AAAS, 1974), pp. 125–136 (rpt. from *Science* [30 December 1955]: 1255–1259; Weaver, "The Encouragement of Science," *Scientific American* 199, no. 3 (September 1958): 170–178; and Weaver, "A Great Age for Science," in *Goals for Americans* (Englewood Cliffs, N.J.: Prentice-Hall, 1960), pp. 103–126 (rpt. as Weaver, *A Great Age for Science* [New York: Sloan Foundation, 1962]); Weaver, "Understanding Science," 12 October 1948, Rockefeller Archive Center, North Tarrytown, NY, Rockefeller Foundation archives [hereafter referred to as RF], RG1.1, ser. 200F, box 175, folder 2124; Weaver diary, RF, RG12, 29 January 1982. (In the indexes to Weaver's diaries, few people receive more than a couple of citations each year; the people who do receive more are often those with whom Weaver discussed his popular-science activities.)

13. Michael M. Sokal, "*Science* and James McKeen Cattell, 1894–1945," *Science* 209 (4 July 1980): 43–52; and John Walsh, "*Science* in Transition," *Science* 209 (4 July 1980): 52–57.

14. Paul Klopsteg to J. W. Buchanan, 20 November 1951, AAAS executive office files, "Arden House" folder.

15. Wolfle, *Renewing a Scientific Society*, pp. 46–47; and A. J. Carlson, A. C. Ivy, and Ralph A. Rohweder to Howard Meyerhoff, 7 June 1951, AAAS executive office files, "Arden House" folder.

16. For introductions to the literature on academics and anticommunism in the 1950s, see Lionel S. Lewis, *Cold War on Campus: A Study of the Politics of Organizational Control* (New Brunswick/Oxford: Transaction Books, 1988) and Ellen W. Schrecker, *No Ivory Tower: McCarthyism and the Universities* (New York/ Oxford: Oxford University Press, 1986); more particularistic studies and commentary can be found in David R. Holmes, *Stalking the Academic Communist: Intellectual Freedom and the Firing of Alex Novikoff* (Hanover/London: University Press of New England, 1989) and I. F. Stone, *The Haunted Fifties, 1953– 1963* (Boston: Little Brown, 1989).

17. Wolfle, *Renewing a Scientific Society*, pp. 48–50; Weaver to Howard Meyerhoff, 30 July 1951, AAAS executive office files, "Arden House" folder; and *Science* 114 (31 August 1951): 246. Weaver was so impressed by Arden House that he recommended its use to Rockefeller Foundation trustees. The conference rooms were "exceedingly well equipped," the food "was excellent but not overly elaborate," and the surroundings "are completely restful and beautiful . . . so well protected that, when driving up to the house, we saw five deer, and a doe with twin fawns sauntering past the dining room window as we were eating dinner." Weaver to C. I. Barnard, 18 September 1951, RF, RG2, ser. 200, box 518, folder 3404.

18. Warren Weaver, "AAAS Policy," *Science* 114 (2 November 1951): 471–472 (emphasis in original); and John Pfeiffer, taped interview, 8 September 1987, New Hope, Penn. (Pfeiffer, a freelance science writer, was one of the consultants at the Arden House meeting). Weaver conveniently neglected to note that the "public understanding" section of the constitution was then less than five years old.

19. Weaver to "Gentlemen," 3 September 1951, AAAS executive office files, "Arden House" folder; AAAS executive committee minutes, 15–16 September 1951, p. 5; and Weaver, "AAAS Policy," 1951. At the time, the "manpower" problem was the hope (on the part of scientists) that the nation's scientists not be drafted indiscriminately, but be detailed to war-related scientific and technological positions that used their special training. Like so many of the issues that concerned scientists in the postwar years, the manpower issue had both altruistic, patriotic components and strong elements of self-interest.

20. Meyerhoff to Warren Weaver, 8 August 1951, AAAS executive office files, "Arden House" folder; and AAAS executive committee minutes, 15–16 September 1951, p. 4.

21. J. C. Jensen to Howard Meyerhoff, 10 October 1951, AAAS executive office files, "Arden House" folder.

22. R. W. Gerard to Howard Meyerhoff, 15 November 1951, AAAS executive office files, "Arden House" folder.

23. Weaver to Howard Meyerhoff, 8 October 1951, AAAS executive office files, "Arden House" folder.
24. John Burnham, *How Superstition Won and Science Lost* (New Brunswick, N.J.: Rutgers University Press, 1987)—I am grateful to Burnham for many years ago sharing his book with me in draft form; Marcel C. LaFollette, *Making Science Our Own: Public Images of Science, 1910–1955* (Chicago: University of Chicago Press, 1990).
25. Relis B. Brown to Howard Meyerhoff, 13 December 1951, and J. W. Buchanan to Paul Klopsteg, 12 November 1951, both in AAAS executive office files, "Arden House" folder.
26. Charles E. Odegaard to Howard Meyerhoff, 30 October 1951, AAAS executive office files, "Arden House" folder. For an analysis of the links between the self-image of scientists in this period and the ideals of democratic society, see David Hollinger, "The Defense of Democracy and Robert K. Merton's Formulation of the Scientific Ethos," *Knowledge and Society* 4 (1983): 1–15.
27. Wolfle, *Renewing a Scientific Society*, esp. pp. 41–69. For records of the period, see the "Arden House" folder in the AAAS executive office files.
28. Wolfle, *Renewing a Scientific Society*, pp. 51–52 and 90–92, contains an illuminating discussion of the personal difficulties between Meyerhoff and other AAAS leaders. See also Ralph A. Rohweder to Weaver, 18 February 1953, AAAS executive office files, "Arden House" folder; and Earl Ubell, "Scientists Out to Give Ideas Popular Appeal: American Association to Reorganize in Face of 'Intellectual Bankruptcy,'" *New York Herald Tribune*, 29 December 1952, p. 15.
29. Interview with Dael Wolfle, 14 February 1987, Chicago, Ill.; Wolfle, *Renewing a Scientific Society*, pp. 90–93; Howard A. Meyerhoff, "Boston, 1953," *Science* 117 (20 February 1953): 3A (also published in *Scientific Monthly* 76 [March 1953]: 195).
30. Wolfle, *Renewing a Scientific Society*, pp. 52–57, 73–74; and John Behnke to Ralph A. Rohweder, 27 January 1953, AAAS executive office files, "Arden House" folder.
31. Kennard Baker Bork, *Cracking Rocks and Defending Democracy: Kirtley Fletcher Mather, Scientist, Teacher, Social Activist, 1888–1978* (San Francisco: AAAS Pacific Division, 1994).
32. Kirtley Mather, "Common Ground of Science and Politics," *Science* 117 (20 February 1953): 169–174; Chester I. Barnard, "Warren Weaver: President-Elect of AAAS," *Science* 117 (20 February 1953): 174–176; and Ralph Rohweder to Weaver, 18 February 1953, AAAS executive office files, "Arden House" folder.
33. AAAS board minutes, December 1953, item 16.
34. AAAS board minutes, 6–7 March 1954, item 17; Wolfle to members of AAAS board of directors, 13 May 1954, AAAS executive office files, "Arden House" folder; and AAAS board minutes, May 1954, item 5. Warren Weaver rejoiced at Wolfle's presence, for he had considered Meyerhoff an obstacle to advancing the public understanding of science plans; Weaver diary, 21 May 1953, RF. On the change in Wolfle's title, see Wolfle, *Renewing a Scientific Society*, pp. 94–95. See also Dael Wolfle, "The Future of the AAAS," *Science* 119 (19 February 1954): 3A.

35. Wolfle, *Renewing a Scientific Society*, pp. 56–57; and Wolfle, personal communication, 16 April 1987.

36. Wolfle, *Renewing a Scientific Society*, pp. 77–78; and Walsh, "*Science* in Transition," 1980, pp. 54–55. For details of these negotiations and related activities, see the records in Borras papers, AAAS archives, box 9, "Scientific American Inc., 1949–1955" folder.

37. Wolfle, *Renewing a Scientific Society*, pp. 77–78; Walsh, "*Science* in Transition," 1980, pp. 54–55; Wolfle interview, 1987; and Gerard Piel, oral history, 23 September 1986, American Philosophical Society, Philadelphia, pp. 59–60.

38. Sokal, "*Science* and Cattell," p. 48; see also Burton E. Livingston to members of AAAS executive committee, 12 October 1942, AAAS archives, F. R. Moulton papers, box 14; and Wolfle, *Renewing a Scientific Society*, pp. 74–75, citing *Scientific Monthly* 63 (December 1948): 488.

39. AAAS board minutes, 3–4 March 1956, addendum no. 4, suppl. to agenda item 30; Graham DuShane, "Report to the editorial board and agenda for meeting, 8 December 1956," 6 December 1956, and "AAAS and American Science. Report of the Committee on Membership Development for the American Association for the Advancement of Science," 7 December 1956, both in AAAS archives, "Scientific Monthly" file; Wolfle, *Renewing a Scientific Society*, pp. 74–75; AAAS board minutes, 6 June 1956, p. 3; Dael Wolfle, "Possible merger of *Science* and *The Scientific Monthly*," 9 October 1956, AAAS board minutes, October 1956, tab D; AAAS board minutes, 27–28 December 1956, item 26; AAAS board minutes, 16 March 1957, item 24, p. 5; AAAS board minutes, October 1957, tab B, p. 7; AAAS board minutes, 21–22 June 1958, p. 4; and Walsh, "*Science* in Transition," pp. 52–53.

40. Unless indicated otherwise, this section on education is drawn almost entirely from Wolfle, *Renewing a Scientific Society*.

41. I'm grateful to longtime *Science* staff writer and editor John Walsh for pointing out, in written comments on an earlier draft of this chapter, the changed attitudes of AAAS leaders.

42. *AAAS Bulletin* 19, no. 2 (June 1974): 15.

43. AAAS board minutes, 3–4 March 1956, p. 4; and "Review of decisions concerning association activities reached at June 1956 board meeting," AAAS board minutes, June 1958, tab A.

44. On the certification studies, see *AAAS Bulletin* 8, no. 3, p. 1.

45. *AAAS Bulletin* 9, no. 1 (March 1964): 3–4.

46. Final quote is from Wolfle, *Renewing a Scientific Society*, p. 168.

47. *Science* 225 (3 August 1984): 496; Asimov letter is in *Science* 140:42.

48. Available at www.sciencenetlinks.com

49. Kirtley Mather, "The Common Ground of Science and Politics," in Kargon, *Maturing of American Science*, pp. 113–123.

50. Mather, "Common Ground," p. 120.

51. Daniel Greenberg, "AD-X2: The Case of the Mysterious Battery Additive Comes to an End," *Science* 134 (29 December 1961): 2086–2087.

52. Lee Lorch et al., "Discriminatory Practices [letter]," *Science* 114 (10 August 1951): 161–162; Council for the Advancement of Negroes in Science, "Memorandum

on the Situation with Regard to Negroes in Science in the U.S.A." [n.d., "submitted to the American Association for the Advancement of Science for consideration at its 1953 annual convention, held in Boston, December 26–31"], AAAS archives (penciled notations indicate the Weaver response and lack of other action).

53. Wolfle, *Renewing a Scientific Society*, pp. 53–55; Carleton Mabee, "Margaret Mead's Approach to Controversial Public Issues: Racial Boycotts in the AAAS," *Historian* 48 (1986): 191–208; and Clifford Grobstein to Dael Wolfle, 27 October 1955; Wolfle to Grobstein, 2 November 1955; and Grobstein to Wolfle, 7 November 1955, all in AAAS archives.

54. "1958 Parliament of Science," *Science* 127 (18 April 1958): 852–858.

55. "1958 Parliament of Science," p. 856.

56. *Science* 125 (1957): 147.

57. AAAS Committee on Science in the Promotion of Human Welfare, "Science and Human Welfare," *Science* 132 (8 July 1960): 68.

58. AAAS Committee on Science in the Promotion of Human Welfare, "Science and Human Survival," *Science* 134 (29 December 1961): 2080–2083.

59. Victor Cohn, "A Science Writer's View," *AAAS Bulletin* 7, no. 1 (January 1962): 5.

60. *AAAS Bulletin* 7, no. 1 (January 1962): 6.

61. Sharon Dunwoody, "The Science Writing Inner Club: A Communication Link Between Science and the Lay Public," *Science, Technology & Human Values* 5 (Winter 1980): 14–22; Christopher Dornan, "The 'Problem' of Science and the Media: A Few Seminal Texts in Their Context, 1956–1965," *Journal of Communication Inquiry* 12, no. 2 (1988): 53–70; Christopher Dornan, "Some Problems in Conceptualizing the Issue of 'Science and the Media,'" *Critical Studies in Mass Communication* 7, no. 1 (1990): 48–71.

62. Alan T. Waterman, "Letter from the President." *AAAS Bulletin* 8, no. 2 (August 1963): 1–2.

63. Details of the book and symposia publication can be followed in the annual reports of the executive officers.

64. Information on the regional divisions comes from Wolfle, *Renewing a Scientific Society*, pp. 3–4; Lora Mangum Shields and Marlowe G. Anderson, "The AAAS Southwestern and Rocky Mountain Division—How a Regional Division Operates," *Science* 194 (24 December 1976): 1407; Karen Hopkin, "Arctic Division Meeting Addresses Environmental Change," *Science* 257 (25 September 1992): 1962; Diana Pabst, "AAAS Pacific Division Meets in British Columbia," *Science* 269 (28 July 1995): 575; the annual *AAAS Handbook*; and annual reports of the executive officers.

65. Waterman, "Letter from the President," pp. 1–2.

66. See the "Arden House" folder, AAAS executive office files, esp. Kirtley Mather to colleagues, 1 July 1952; Warren Weaver to Mather, 31 July 1952; and Howard Meyerhoff to Weaver, 7 August 1952; and AAAS board minutes, October 1958.

67. Sherburne to AAAS Committee on Public Understanding of Science, 12 May 1961, pp. 5–7, and Minutes of AAAS Committee on Public Understanding of Science, 31 July 1961, p. 4, both in "Public Understanding of Science, 1958–

1964" folder; "Status Report, Public Understanding of Science Department," AAAS board minutes, March 1962, pp. 1, 3; Wolfle, *Renewing a Scientific Society*, pp. 189–210; "Sherburne to head new AAAS program to improve public understanding," *Science* 132 (16 December 1960): 1823–1824; and E. G. Sherburne, Jr., interview, 9 March 1987, Washington, D.C.

68. Wolfle, *Renewing a Scientific Society*, pp. 189–210; E. G. Sherburne, "A Project to Improve the Quality and Quantity of Science Programming on Television: Final Report," 15 October 1965, "Public Understanding of Science, 1958–1964" folder; "Status Report, Public Understanding of Science Department," AAAS board minutes, March 1962, tab D, p. 5; "Status Report, Public Understanding of Science Program," AAAS board minutes, March 1964, tab E, p. 6; AAAS board minutes, June 1964, tab A, p. 2; Wilbur Schramm, *Science and the Public Mind* (Stanford: Institute for Communication Research, 1962; distributed by AAAS); and Philip Tichenor, Ph.D. diss., Stanford University, 1963.

69. "Status Report, Public Understanding of Science Program," AAAS Board, March 1964, tab E, pp. 2–3.

70. Rachel Carson, *Silent Spring* (Boston: Houghton Mifflin, 1962); correspondence on the founding of SIPI can be found in RF 261.2, ser. 200, box 85, folder 732; and Walter McDougall, . . . *The Heavens and the Earth* (New York: Basic Books, 1985), p. 389. Setting this scene requires a complex history of the early 1960s. Two contemporary sources are Barry Commoner, *Science and Survival* (New York: Viking Press, 1966); and Gerald Holton, ed., *Science and Culture* (Boston: Beacon Press, 1965). More historically analytical are Siegel, *Troubled Journey*, 1984; Frank Graham, Jr., *Since Silent Spring* (Boston: Houghton Mifflin, 1970); and Milton S. Katz, *Ban the Bomb: A History of SANE, the Committee for a Safe Nuclear Policy, 1957–1985* (New York/Westport, Conn./London: Greenwood Press, 1987).

71. Polykarp Kusch to Gordon MacDonald and Thomas Park, 10 August 1964; Kusch to Dael Wolfle, 7 August 1964; and Wolfle to Kusch, 19 August 1964, all in "Public Understanding of Science, 1958–1965" folder.

72. Henry Guerlac, "Some Scientific Plain Talk," *New York Times*, 27 December 1953, sec. 7, p. 11.

73. Frederick Seitz to Philip Abelson, 20 November 1963, Borras papers, box 10, "*Science* and *The Scientific Monthly* (old, I)" folder.

74. Dael Wolfle to AAAS board, 13 December 1963, and Wolfle to AAAS board, 4 February 1964, both in Borras papers, box 10, "*Science* and *The Scientific Monthly* (old, I)" folder.

75. Minutes of the meeting of the AAAS Committee on Public Understanding of Science, 5 October 1963, AAAS archives, "*Science and Public Policy*" folder, item 11 and appendix A; Wolfle interview, 14 February 1987.

76. Weaver to Dael Wolfle, 20 September 1963, AAAS archives, "*Science and Public Policy*" folder.

77. Daniel S. Greenberg to Dael Wolfle, 12 June 1964, and Wolfle to AAAS board and Committee on Publications, 1 May 1964, both in AAAS archives, "*Science and Public Policy*" folder; and AAAS board minutes, June 1964, p. 6.

78. "Manpower" estimates rely on complex surveys and statistics, making cross-

time comparisons difficult to interpret. These numbers come from NSF, *National Patterns of R&D Resources: Funds & Manpower in the United States, 1953–1973,* NSF 73–303 (Washington, D.C.: NSF, 1973) and National Science Foundation, *National Patterns of R&D Resources: 1996,* NSF 96–333 (Arlington, Va.: NSF, 1996).

79. Alfred Romer, "Letter from the President," *AAAS Bulletin* 11, no. 3 (September 1966): 2; John Mayor, "Farewell," *AAAS Bulletin* 15, no. 2 (June 1970): 1. I want to add my own agreement to the assessments above. In the mid-1980s, as I first began working on AAAS history, Wolfle kindly sat with me for interviews and allowed me to read his memoirs in manuscript. I found him a tremendously gracious and thoughtful person.

80. Wolfle, *Renewing a Scientific Society*, p. 95; Daniel Koshland, "Philip Hauge Abelson," *Science* 227 (4 January 1985), editorial.

81. Allen Hammond, in Philip H. Abelson, *Enough of Pessimism: 100 Essays* (Washington, D.C.: AAAS, 1985).

82. Daniel Greenberg, *The Politics of Pure Science* (New York: New American Library, 1967), p. 209.

83. For example: Allen Hammond, William Metz, and Thomas Maugh, *Energy and the Future* (1973); Jean L. Marx and Gina Bari Kolata, *Combating the #1 Killer* (1978); Philip H. Abelson, ed., *Health Care: Regulation, Economics, Ethics, and Practice* (1978). *Science*'s news staff also provided a training ground for other media; by the late 1990s, at least two dozen prominent science journalists at outlets from the *New York Times*, the *Wall Street Journal*, and National Public Radio to the *Economist* and NBC were *Science* alumni.

84. "Science: Report to the Board of Directors," *AAAS Bulletin* 8, no. 2 (1963), p. 4.

85. Ibid.

86. F. J. Vine, "Spreading of the Ocean Floor: New Evidence." *Science* 154 (16 December 1966): 1405–1415; Garrett Hardin, "The Tragedy of the Commons." *Science* 162 (13 December 1968): 1243–1248.

87. *AAAS Bulletin* 15, no. 1 (March 1970): 3.

88. Frederick Mosteller, "The Next 100 Years of Science," *Science* 209 (4 July 1980): 21.

89. W. D. Carey, "Science: Its Place in AAAS," *Science* 209 (4 July 1980): 24–26.

90. Biographical information on Golden appears in *Who's Who in America*. Among his books are several edited volumes exploring scientific advice to governments; the most recent include William T. Golden, ed., *Science and Technology Advice to the President, Congress, and Judiciary,* 2nd ed. (Washington, D.C.: AAAS Press, 1993) and William T. Golden, ed., *Worldwide Science and Technology Advice to the Highest Levels of Governments* (New York: Pergamon, 1991).

91. *AAAS Bulletin* 16, no. 2 (June 1971): 1–2; Wolfle, *Renewing a Scientific Society*, pp. 179–181; Mina Rees and William Bevan, "Report to the Association—1972," *Science* 179 (23 February 1973): 824; William Bevan, "Report to the Association—1973," *Science* 184 (26 April 1974): 497.

92. Wolfle, *Renewing a Scientific Society*, pp. 179–181.

93. *AAAS Bulletin* 7, no. 3 (September 1962).

94. Walter Orr Roberts, "President's Letter," *AAAS Bulletin* 13, no. 3 (1968): 1.
95. Bentley Glass, "President's Letter," *AAAS Bulletin* 14, no. 3 (1969): 1–2.
96. *AAAS Bulletin* 11, no. 4 (1966): 12.
97. Data on R&D expenditures comes from the AAAS R&D Budget and Policy Project, especially from data and charts available online in March 1998 at www.aaas.org/spp/dspp/rd/guide.htm#total.
98. *AAAS Bulletin* 15, no. 1 (1970): 1; James K. Glassman, "AAAS Boston Meeting: Dissenters Find a Forum," *Science* 167 (2 January 1970): 36–38.
99. Glassman, "AAAS Boston Meeting," pp. 36–38.
100. *AAAS Bulletin* 15, no. 1 (1970): 2.
101. Ibid.: 1.
102. Wolfle, *Renewing a Scientific Society*, pp. 247, 251–252; Jeffrey K. Stine, *Twenty Years of Science in the Public Interest: A History of the Congressional Science and Engineering Fellowship Program* (Washington, D.C.: AAAS, 1994), pp. 8–9; John Walsh, "Science for the People: Comes the Evolution," *Science* 191 (12 March 1976): 1033–1035; Donna Jeanne Haraway, "The Transformation of the Left in Science: Radical Associations in Britain in the 30's and the U.S.A. in the 60's," *Soundings* 58 (Winter 1975): 441–462.
103. Wolfle, *Renewing a Scientific Society*, pp. 140–142.
104. "William Bevan," *American Psychologist* 45 (April 1990): 480.
105. William Bevan, "The General Scientific Association: A Bridge to Society at Large," *Science* 172 (23 April 1971): 349–352.
106. Gerard Piel, "1970 Report of the Committee on Public Understanding of Science," 11 December 1970; and James C. Butler and William Bevan, "A Report to the Alfred P. Sloan Foundation . . . ," n.d., "Public Understanding of Science, 1972–1973" folder. The studies, financed with funds remaining from the money given to the AAAS by the Sloan Foundation in the early 1960s, were published as: David Prowitt, "Science Programming on Radio and Television," AAAS Misc. Publ. 72–17, September 1972; F. James Rutherford, "Students and Science," AAAS Misc. Publ. 72–15, August 1972; Arthur H. Livermore, "Science Kits," AAAS Misc. Publ. 72–16, July 1972; Richard K. Winslow, "Publishing for Scientific Literacy," AAAS Misc. Publ. 72–11, May 1972; and Park Teter, "Education for Self-Reliance," AAAS Misc. Publ. 72–7, March 1972; copies of the reports are in the "Public Understanding of Science–Reports, 1970–1980" folder.
107. "Appendix I: Communications Programs in the Public Understanding of Science and Technology," proposal to the NSF, n.d., p. 7; James C. Butler to William Bevan, 14 September 1973; and James C. Butler and William Bevan, "A Report to the Alfred P. Sloan Foundation . . . ," n.d., all in "Public Understanding of Science, 1972–1973" folder.
108. "Appendix I: Communications Programs in the Public Understanding of Science and Technology," proposal to the NSF, n.d.; Gerard Piel, "1972 Report of the AAAS Committee on the Public Understanding of Science," n.d.; "Preliminary working paper for Decision Maker seminar," September 1972; and prospectus, "Science and the Media," draft dated 14 September 1972, all in "Public Understanding of Science, 1972–1973" folder; and "Summary Report, 1976

Mass Media Intern Program," n.d., in "Public Understanding of Science—Reports, 1970–1980" folder; James C. Butler, "1973 Report of the AAAS Committee on the Public Understanding of Science," n.d., in "Public Understanding of Science, 1972–1973" folder.

109. Mafalda Marrocco, "Science Bids for Prime-Time Television Audience," *AAAS Bulletin* 18, no. 5 (November 1973): 1–3; Mafalda Marrocco, "NOVA Ends First Season," *AAAS Bulletin* 19, no. 2 (June 1974): 14.

110. Stine, *Twenty Years of Science in the Public Interest*.

111. Bruce V. Lewenstein, *A Brief History of the AAAS Mass Media Program* (Washington, D.C.: AAAS, 1995).

112. For analysis of the "manpower" studies, see Margaret Rossiter, *Women Scientists in America: Before Affirmative Action, 1940–1972* (Baltimore: Johns Hopkins University Press, 1995).

113. Wolfle, *Renewing a Scientific Society*, pp. 25–26; Haraway, "Transformation of the Left in Science," pp. 457–458.

114. *AAAS Bulletin* 18, no. 5 (November 1973): 5.

115. "Office of Opportunities in Science," *Science* 186 (6 December 1974): 919, 949; "Office of Opportunities in Science," *Science* 188 (2 May 1975): 439–440; "Native American Contributions to Science, Engineering, and Medicine," *Science* 189 (4 July 1975): 38–39. See also the annual reports of the executive officer published in *Science*.

116. J[ohn] W[alsh], "Rocky Speaks at AAAS Meeting," *Science* 191 (5 March 1976): 927; "Disabled Scientists Featured in *Newsweek, Atlantic*," *Science* 194 (3 December 1976): 1038; "Handicapped Resource Group Members Work for Barrier Elimination," *Science* 195 (4 February 1977): 475–476. The "huge" quote is from William D. Carey, "1979 Report of the Executive Officer," *Science* 207 (22 February 1980): 865.

117. Shirley M. Malcom, "The Audiences to Listen To," in *When Science Meets the Public*, ed. Bruce V. Lewenstein (Washington, D.C.: AAAS, 1992), pp. 148–150.

118. *AAAS Bulletin* 8, no. 3, p. 4.

119. "'Science and Man in the Americas' Attracts 5,000 Participants," *AAAS Bulletin* 18, no. 4 (September 1973): 17. One of the organizers of the Mexico City meeting was Albert Baez, chairman of the AAAS's Committee on Scientific Education and father of the famous folk singer Joan Baez (*AAAS Bulletin* 16, no. 2 [June 1971]: 3).

120. Information on international programs drawn from various articles in *AAAS Bulletin*, the "AAAS News" section of *Science*, annual reports of the executive officer, and the *AAAS Handbook*. See in particular *AAAS Bulletin* 18, no. 6 (December/January 1974): 8.

121. Helen Seaborg, who had been secretary to Glenn Seaborg's mentor E. O. Lawrence, also a Nobelist, once noted that she may be the only non-Nobelist in history who has twice answered the phone call from Stockholm.

122. Roger Revelle, "The Scientist and the Politician," *Science* 187 (21 March 1975): 1100–1105; Margaret Mead, "Towards a Human Science," *Science* 191 (3 March 1976): 903–909; Emilio Q. Daddario, "Science and Its Place in Society," *Science* 200 (21 April 1978): 263–265.

123. Anna J. Harrison, "Reflections on Current Issues in Science and Technology," *Science* 215 (26 February 1982): 1061–1063. For examples of more traditional biographical sketches, see Louis Levin, "William D. McElroy, President-Elect," *Science* 187 (21 March 1975): 1097–1100; Larry D. Singell, "Kenneth E. Boulding, President-Elect of the AAAS," *Science* 200 (21 April 1978): 289–290; William Kruskal, "Frederick Mosteller, President-Elect," *Science* 203 (2 March 1979): 866–867; and Joseph Weneser, "D. Allan Bromley, President-Elect," *Science* 207 (22 February 1980): 861–862.
124. Wolfle, *Renewing a Scientific Society*, pp. 234–235.
125. Philip M. Boffey, "Dispute over Jensen Election as Fellow Flares in Council," *Science* 195 (11 March 1977): 965; Mabee, "Margaret Mead's Approach to Controversial Public Issues."
126. William D. Carey, "Moratorium on Nomination of AAAS Fellows," *Science* 200 (7 April 1978): 39–40; Catherine Borras, "AAAS Council Meeting, 1979," *Science* 203 (2 March 1979): 871–874, on p. 872; "Nomination of AAAS Fellows Invited," *Science* 203 (9 March 1979): 996, 1034.
127. Wolfle, *Renewing a Scientific Society*, pp. 53–55; Catherine Borras, "AAAS Council Meeting, 1975," *Science* 187 (21 March 1975): 1114–1117, on p. 1116; "'79 AAAS Meeting Moved from Chicago to Houston," *Science* 199 (3 March 1978): 954; the details of the 2003 Denver meeting have been relayed to me by Michael Spinella and Ed Leonardo of the AAAS meetings office.
128. *AAAS Bulletin* 17, no. 2 (1972): 4.
129. Athelstan Spilhaus, "President's Letter," *AAAS Bulletin* 15, no. 3 (September 1970): 2.
130. See Wolfle, *Renewing a Scientific Society*, ch. 7, for more information, esp. pp. 134–135.
131. *AAAS Bulletin* 12, no. 3 (September 1967): 1.
132. Numbers drawn from the *AAAS Handbook* show the number of members who join a section and those who do not; percentages are based on total individual membership in AAAS.
    1972: 118,900 (93%)/8500 (7%)
    1974: 66,700 (54%)/ 56,800 (46%)
    1985: 90,200 (67%)/44,700 (33%)
    1997: 75,000 (53%)/ 66,600 (47%)
133. Philip H. Abelson, "AAAS Meetings," *Science* 183 (1 February 1974): 369; William D. Carey, "The Denver Meeting: Afterthoughts," *Science* 196 (15 April 1977): 265.
134. Talk titles taken from annual meeting programs.
135. Details of the financial problems can be followed in annual reports: Mina Rees and William Bevan, "Report to the Association—1972," *Science* 179 (23 February 1973): 824–827; William Bevan, "Report to the Association—1973," *Science* 184 (26 April 1974): 496–500; Leonard M. Rieser, "Report from the Retiring Chairman," *Science* 197 (21 March 1975): 1106–1108; Philip M. Abelson, "Report to the Association," *Science* 197 (21 March 1975): 1109–1114.
136. John Walsh, "Carey to be AAAS Executive Officer," *Science* 186 (15 November 1974): 611.

137. William D. Carey, "Setting Priorities: The Slippery Slope," *Vital Speeches of the Day* 52 (September 1986): 732–734; "Scientific Exchanges and U.S. National Security," *Science* 215 (8 January 1982): 139–141.

138. William D. Carey, "Foreword," in Willis H. Shapley, *Research and Development in the Federal Budget: FY 1977* (Washington, D.C.: AAAS, 1976). Shapley was the son of 1947 AAAS president Harlow Shapley, and his daughter Deborah worked for *Science* for some years as a reporter.

139. Discussions of the planned magazine can be found in AAAS board minutes from April 1978 forward. A national popular science magazine had been one of James Butler's ambitious ideas, present in both his earliest surveys of AAAS activities and in his 1971 proposal to the National Science Foundation. Hammond's plans, however, appear to have been developed entirely independently of Butler's original ideas; see James C. Butler, "A Presentation to the Committee on the Public Understanding of Science," n.d., p. 8; and "Appendix I: Communications Programs in the Public Understanding of Science and Technology," proposal to the NSF, n.d., p. 25.

140. AAAS board minutes, 20–21 April 1979, pp. 4, 6.

141. Bruce V. Lewenstein, "Was There Really a Popular Science 'Boom'?" *Science, Technology & Human Values* 12, no. 2 (spring 1987): 29–41.

142. Luis W. Alvarez, Walter Alvarez, Frank Asaro, and Helen V. Michel, "Extraterrestrial Cause for the Cretaceous-Tertiary Extinction," *Science* 208 (6 June 1980): 1095ff; Jean L. Marx, "New Disease Baffles Medical Community," *Science* 217 (13 August 1982): 618–621. See also Ruth M. Kulstead, "Publishing AIDS Papers in the early 1980s," in Caroline Hannaway, Victoria A. Harden, and John Parascandola, eds., *AIDS and the Public Debate: Historical and Contemporary Perspectives* (Amsterdam: IOS Press, 1995), pp. 107–123.

143. John T. Edsall, *Scientific Freedom and Responsibility: A Report of the AAAS Committee on Scientific Freedom and Responsibility* (Washington, D.C.: AAAS, 1975), p. x.

144. Edsall, *Scientific Freedom and Responsibility*, p. 45.

145. See, for example, Clyde Collins Snow, Eric Stover, and Kari Hannibal, "Scientists as Detectives: Investigating Human Rights," *Technology Review* 92 (February/March 1989): 42–49; Clyde Snow, "Murder Most Foul," *Sciences* 35 (May/June 1995): 16–20; and Christopher Joyce and Eric Stover, *Witnesses from the Grave: The Stories Bones Tell* (Boston: Little, Brown and Company, 1991).

146. Floyd Bloom, "The Koshland Years—A Decade of Progress," *Science* 268 (5 May 1995): 619; "Koshland Named *Science* Editor," *Science* (17 Aug 1984): 695.

147. Daniel E. Koshland, Jr., "A New Look," *Science* 231 (3 January 1986): 9.

148. Bloom, "The Koshland Years," p. 619, covers the international expansion. The ending of the relationship with Scherago and its effects are my own interpretations of undocumented conversations I had with AAAS and *Science* staff during the early 1990s.

149. Julia King, "Are Scientific Societies Diverted by Their Cash Cows?" *Scientist* 5, no. 10 (13 May 1991): 1, 10, 25.

150. Figures reported in AAAS annual reports indicate that *Science* regularly returns

to the association (through subscription income and advertising revenue) almost the same amount of money that it receives for operating costs. Most of the subscription income comes from the annual amounts paid by AAAS members, which are officially divided into "*Science* subscription" and "association dues" categories. However, the number of people who would join the association if that were a choice independent of *Science* subscription is not known.

151. R. K. Saiki et al., "Enzymatic Amplification of Beta-Globin Genomic Sequences and Restriction Site Analysis for Diagnosis of Sickle Cell Anemia," *Science* 230 (20 December 1985): 1350–1354 (coauthor Kary Mullis would win the Nobel Prize for his discovery of PCR; see Paul Rabinow, *Making PCR: A Story of Biotechnology* [Chicago: University of Chicago Press, 1996] for discussion of the publication history); J. R. Riordan, J. M. Rommens, and B. Kerem, "Identification of the Cystic Fibrosis Gene: Cloning and Characterization of Complementary DNA," *Science* 245 (8 September 1989): 1066–1073; B. Kerem, J. M. Rommens, and J. A. Buchanan, "Identification of the Cystic Fibrosis Gene: Genetic Analysis," *Science* 245 (8 September 1989): 1073–1080; Bloom, "The Koshland Years," 619; Karen Hopkin, "AAAS Purchases Land for New Headquarters," *Science* 258 (18 December 1992): 1962.

152. Institute for Scientific Information, *1996 Science Citation Index Journal Citation Reports* (Philadelphia: ISI, 1997).

153. Some examples include James Butler, Richard Scribner, William Wells, Al Teich, Shirley Malcom, Norman Metzger, and Carol Rogers.

154. Colleen Cordes, "Selection of a Low-Key Official at NSF to Lead Science Group Draws Praise, Raises Concern," *Chronicle of Higher Education*, 8 February 1989, pp. A17, A24.

155. Project 2061, *Science for All Americans* (Washington, D.C.: AAAS, 1989); AAAS, *Benchmarks for Science Literacy* (New York: Oxford University Press, 1993).

156. Karen Hopkin, "AAAS Purchases Land for New Headquarters," *Science* 258 (18 December 1992): 1962; Cynthia Lollar, "AAAS to Build New Headquarters," *Science* 262 (26 November 1993): 1456–1457; Diana Pabst, "Building for the Future," *Science* 267 (24 February 1995): 1200–1201; Cynthia Lollar, "AAAS Dedicates Golden Center for Science," *Science* 277 (26 September 1997): 2012.

157. Sheila Jasanoff et al., "Conversations with the Community: AAAS at the Millenium," *Science* 278 (19 December 1997): 2066–2067.

158. Paul R. Gross, "Response to 'Conversations with the Community . . . ,'" electronic message posted 2 January 1998 (www.nextwave.org/server-java/Forum/recruit/msg-883721024177).

About the Contributors

Sally Gregory Kohlstedt has been professor in the History of Science and Technology Program at the University of Minnesota since 1989. At the time of writing, she is director of the Center for Advanced Feminist Studies as well. She received her undergraduate degree from Valparaiso University and her Ph.D. from the University of Illinois. She has taught at Simmons College in Boston and at Syracuse University. She has been a visiting professor at the University of Melbourne, Australia, and the Amerika Institut at the University of Munich, Germany. She has held fellowships from the Smithsonian Institution, the Woodrow Wilson Center, and the American Antiquarian Society, among others. She has published numerous articles; presented her research results at universities, museums, and before public audiences; and served on review boards in the sciences, social sciences, and humanities.

Her professional contributions include numerous offices and she has chaired several committees within the History of Science Society, including service as secretary from 1978 to 1981, and president in 1992 and 1993. She has also been active in the American Association for the Advancement of Science, elected a Fellow, and twice served on the nominating committee and then as chair of Section L on the History and Philosophy of Science; collaborated with program staff and committees on specific projects; and currently participates as a member of the board of directors, 1998–2001.

Her earliest work was on the history of mid-nineteenth-century science, including *The Formation of the American Scientific Community: The American Association for the Advancement of Science,* 1848–1860 (Urbana: University of Illinois, 1976). She edited (with Margaret Rossiter) a milestone volume on *Historical Writing on American Science* as *Osiris,* 2nd series, 1I (1985). Other volumes relating to her interest in scientific organizations include (with R. W. Home) *International Science and National Scientific Identity: Australia between Britain and the Americas* (Holland: Kluewer Academic Publishers, 1991), and *The Origins of Natural Science in the United States: The Essays of George Brown Goode* (Washington: Smithsonian Institution Press, 1991). More recently she has been working in the related area of women's history and has edited (with Barbara Laslett, Helen Longino, and Evelyn Hammonds), *Gender and Scientific Authority* (Chicago: University of Chicago Press, 1996), (with Helen Longino) *Women, Gender and Science: New Directions* as *Osiris,* 2nd series, 12 (1997), and *Women in the History of Science: Readings from Isis* (Chicago: University of Chicago Press, 1999).

Michael M. Sokal is professor of history at Worcester Polytechnic Institute, Worcester, Massachusetts. At WPI he has been honored with the Trustees' Award for Outstanding Creative Scholarship, the President's Award for Outstanding Project Advising, and a two-year term as Paris Fletcher Distinguished Professor in the Humanities. He is a Fellow of the AAAS, and currently serves as immediate past chair of the AAAS Section on History and Philosophy of Science.

From 1988 through 1992, he served as the first executive secretary of the History of Science Society, editing the *HSS Newsletter* and providing oversight for all other HSS Programs. In 1995, he served as visiting program officer at the National Endowment for the Humanities, where he coordinated a joint NEH–National Science Foundation grants competition, "Science and Humanities: Integrating Undergraduate Education." In August 1998, he began a two-year term as director of NSF's Program in Science & Technology Studies, overseeing all NSF grantmaking for research and training in history, philosophy, and social studies of science and technology.

Sokal was trained as an electrical engineer (Bachelor of Engineering, 1966, The Cooper Union, New York City) and worked as an electronics designer for IBM and NASA in the mid 1960s. He earned his M.A. (1968) and Ph.D. (1972) in the history of science and technology at Case Western Reserve University (Cleveland, Ohio). He has taught at WPI since 1970, has held a Smithsonian Institution postdoctoral fellowship (1973–74) at (what was then) the National Museum of History and Technology, and has taught courses at the College of the Holy Cross and Clark University in Worcester.

He has written extensively on the history of psychology and the history of American science; his articles on these subjects have appeared in the *American Historical Review, American Psychologist, American Scientist, Proceedings of the American Philosophical Society, Science,* and many other journals. His edited books include: *An Education in Psychology: James McKeen Cattell's Journal and Letters from Germany and England, 1880–1888* (MIT Press, 1981); (with Patrice A. Rafail) *A Guide to Manuscript Collections in the History of Psychol-*

ogy and Related Areas (Kraus, 1982); and *Psychological Testing and American Society, 1890–1930* (Rutgers University Press, 1987). He currently edits *History of Psychology*, a scholarly journal launched in 1998 and published quarterly by the American Psychological Association for its Division of the History of Psychology.

Bruce V. Lewenstein is associate professor in the Departments of Communication and Science & Technology Studies at Cornell University in Ithaca, New York. After working as a science writer, he earned his Ph.D. in 1987 at the University of Pennsylvania, jointly in history and sociology of science, and in science and technology policy. His work focuses on historical and contemporary issues involving the public understanding of science. He is a past member of the AAAS committee on public understanding of science and technology, and is on the advisory board to the Sciencenter, an interactive science museum in Ithaca. His work has been supported by the National Science Foundation, the U.S. Department of Agriculture, NASA, and the Chemical Heritage Foundation.

Lewenstein has published an edited book (*When Science Meets the Public* [Washington, D.C.: AAAS, 1992]) and is the editor of the journal *Public Understanding of Science*. He has published many articles and book chapters about the popularization of science and science communication, especially the links among science journalism, science museums, and various science outreach activities. Since 1994, he has worked with Cornell's Laboratory of Ornithology to help develop and evaluate several "citizen science" projects that engage the public in real scientific work. A member of a variety of professional, historical, and scientific organizations, he has presented material on popularization of science throughout the world, including lectures in Indonesia, Australia, Canada, Brazil, and several European countries. He has taught in courses at University Pompeu Fabra in Spain, at the University of Sydney in Australia, and in seminars for journalists in Singapore and Uruguay.

Several times, Lewenstein has directed the creation of archives of contemporary issues in science. Most notably, he helped create the National Science Foundation–funded Cornell Cold Fusion Archive, which provides a central location for historians and sociologists trying to understand the scientific and public response to the saga of cold fusion. He has also codirected NSF-funded archives on the use of DNA fingerprinting in the O. J. Simpson murder trial and on the presence of "Y2K" issues in television broadcasts.

INDEX

presidents, AAAS. *See also* leadership:
addresses of outgoing, 18–19; at
annual meetings, 177–181; terms for,
101
press coverage of annual meetings, 35
press room, AAAS's annual, 127
Press Service, AAAS, 84, 90
Price, Don K., 137, 148
Priestley, Joseph, 43
Primack, Joel, 141
priorities, controversy over, 113–115. *See
also* goals, AAAS
Proceedings, 16, 31: Coast Survey report
in, 30; costs of, 36; in 1890s, 46;
hospitality writings in, 15; lateness of,
65; papers published in, 20, 21;
publication of, 55; purpose of, 9; and
F. W. Putnam, 39, 40; sales decline in,
125–126; section reports in, 187–
188n. 84; *Summary*, 65
professionalism, ideals of, 14
professionalization: of science, 51;
tension surrounding, 19
profits, of *Science* magazine, 83, 158
Progressive Era, 3, 62
Project 2061, 136, 160
Project Physics, 160
Providence, R.I., annual meeting in, 177
psychological laboratories, 50
Psychological Review, 53
psychology: "Current Notes" on, 54; as
tool for torture, 155
public: and diffusion of science, 34; and
geologic science, 10; interest in
science of, 73; outreach programs for,
83–87, 107 (*see also* outreach
programs); support for science of, 9
(*see also* popular science publications)
control of, 48; in 1970s, 126; and
scientific status, 18 (*see also* specific
publications)
public forum, AAAS as, 7, 8, 28
publicity, director of, 84
public policy: attempts to influence, 85;
and science, 114 (*see also* politics)
Public Science Day, 151, 161, 163
public television, 140

publishing, technical, 157
Pulitzer Prize, 84
Pupin, Michael I., 79, 83, 85
Putnam, Frederic Ward, 170: democratic
viewpoint of, 46, 55; paper submitted
by, 26; secretarial appointment of, 39–
41; specialization movement and, 43;
support of women by, 24

quantum notions, 59
quantum physics, 74

race: and AAAS membership, 21;
"scientific" views of, 16
racial discrimination, 146
racial segregation, 121, 122
radio, broadcast: of New York Philharmonic, 107; and public interest in
science, 73; science programs on, 91
Ratchford, J. Thomas, 144, 171
rationalism, 2
recombinant DNA, moratorium on
experiments with, 155
Red-baiting, 118
redefinition, process of, 93–94
Redfield, William, 14
Red Scare, 77
Rees, Mina, 139, 145
Reingold, Nathan, 19
Reinhold, Robert, 138
relativity: complexities of, 74; theory of,
59
religion: expression of opinions on, 45;
science and, 27
reporters, science, 125. *See also* media
Reports, AAGN's, 11
research: AAAS fund for, 37–38; AAAS
goal for, 61; "abstract," 14, 20; basic,
111, 137; federal funding for, 103;
integrity of, 156
research and development: expenditures
for, 211n. 97; support for, 137
researchers, advancement of professional,
61
Revelle, Roger, 145, 149
Richmond, Va., annual meeting in, 179
Rieser, Leonard, 143–144

values, development of, 6
value systems, 112
Vaux, William S., 171
Vienna Congress Rules, 60
Vietnam War, 137
vivisection, 46

Walcott, Charles D., 64, 76
Wallace, William D., 146
war, science and, 3, 4, 124
Ward, Henry B., 170: AAAS activity of,
 62; executive committee appointment
 of, 76; public works project of, 94;
 recruitment efforts of, 89; *Science*
 magazine and, 92; secretarial appointment of, 88
Washington, D.C.: AAAS relocation in,
 156, 157, 162; annual meetings in,
 106, 177, 178, 179, 180, 181; as
 scientific center, 74
Washington Post, 124–125, 139
Waterman, Alan, 125, 126
Watson, Miss C. A., 40
Weaver, Warren: administrative turmoil
 concerning, 114, 206n. 34; on Arden
 House, 205n. 17; desegregation efforts
 of, 122; funding efforts by, 118;
 presidential term of, 123; public
 outreach efforts of, ix, 107, 109, 110,
 111, 127, 204n. 12; *Science* magazine
 and, 129–130
websites, 120, 162
Welch, William H., 64
Westinghouse Science Writing Awards, 91
Whewell, William, 18
Whitney, Willis R., 68
Widnall, Sheila, 149
Wilder, Burton G., 26
Wilkes, Charles, 8

Wilson, Edmund B., 67, 89
Wilson, Pres. Woodrow, 69
Wirt, John L., 171
Wolfle, Dael, 170, 171: democratic
 viewpoint of, 129, 148; initiatives of,
 117; leadership vision of, 131;
 memoirs of, 210n. 79; organizational
 turmoil involving, 206n. 34; public
 outreach efforts of, 127; resignation
 of, 139, 160; and *Science*, 116;
 secretarial appointment of, 106, 114,
 115
women: in AAAS leadership, 145; in
 membership, 21, 24; participation in
 science of, 141–142; in scientific
 societies, 63
Woodley, Sam, 75, 89, 171
Woodward, Robert S., 53, 72, 171
World's Fair Century of Progress (1933),
 90
World War I: AAAS response to, 71–73;
 impact on science of, 3
World War II: American science following, 103; and scientific community, 99
Wrather, William E., 171
Wyman, Jeffries, 171

Xerox Corporation, 135

Youmans, Edward Livingston, 34, 61
youth, participation in science of, 141–
 142
Youth Council, 141

Znaniye (All-Union Knowledge Society),
 144
zoologists, 33
zoology, and evolutionary theory, 26